Paolo Cappabianca
Alessandra Alfieri
Enrico de Divitiis
Manfred Tschabitscher

Atlas of
Endoscopic Anatomy
for Endonasal
Intracranial Surgery

Springer-Verlag Wien GmbH

Dr. Paolo Cappabianca
Dr. Alessandro Alfieri
Prof. Dr. Enrico de Divitiis
Department of Neurosurgery,
University "Frederico II", Naples,
Italy

Prof. Dr. Manfred Tschabitscher
Department of Anatomy,
University of Vienna,
Austria

This work is subject to copyright.
All rights are reserved, whether the whole or part of the material is concerned,
specifically those of translation, reprinting, re-use of illustrations, broadcasting,
reproduction by photocopying machines or similar means,
and storage in data banks.

Product Liability: The publisher can give no guarantee for all the information contained in this book.
This does also refer to information about drug dosage and application thereof. In every individual case the respec-
tive user must check its accuracy by consulting other pharmaceutical literature.
The use of registered names, trademarks, etc. in this publication does not imply, even in the absence
of a specific statement, that such names are exempt from the relevant protective laws and
regulations and therefore free for general use.

© 2001 Springer-Verlag Wien

Originally published by Springer-Verlag Wien New York in 2001
Softcover reprint of the hardcover 1st edition 2001
Typesetting: Stefan Kindel, D-67487 Maikammer

SPIN: 10778118

With 192 partly coloured Figures

CIP data applied for

ISBN 978-3-7091-7255-1   ISBN 978-3-7091-6187-6 (eBook)
DOI 10.1007/978-3-7091-6187-6

Foreword

I am honored to write a foreword for the book that Cappabianca et al. are producing for endoscopic transsphenoidal surgery. Since transsphenoidal pituitary surgery was first developed in the early part of the twentieth century, the surgical technique has evolved continuously. The introduction of the operating microscope into transsphenoidal surgery has brought a new epoch in pituitary surgery for decades. The surgical outcome improved and morbidity has been reduced in pituitary surgery. Now, we are entering a new era when computer-assisted imaging is going to play a dominant role in neurosurgical practice. Surgeons will no longer operate on a patient under direct visual perception. Rather, surgical procedures will be performed under the images produced on display monitors or screens. Many different images may be displayed simultaneously on multiple monitors. Surgical images from different perspectives, images displaying realtime surgical guidance, and images assisted by robotic reinforcement are just a few to mention. Endoscopic images will be one of those types of images that surgeons will operate under.

Although endoscope has been used in neurosurgical practice for almost a century, its use has been generally confined to the cerebral ventricular system, mainly due to the basic requirement of endoscopy that demands a hollow cavity for an operation. It is only recently that the use of endoscope in neurosurgery has been extended beyond the ventricular system. Rhinologic application of endoscopy into the paranasal sinus had shed a light on endonasal neuroendoscopy when an endonasal surgical corridor can be utilized in neurosurgical practice. It is just the beginning for neurosurgeons in their exploration of surgical ventures through the endonasal route utilizing endoscopic visualization.

Microscopic anatomical studies have been reported to assist surgeons for microscopic transsphenoidal surgery. Although the real surgical anatomy is basically the same, the optical distortion of endoscopic images is quite substantial when they are compared to microscopic depictions. An endoscope lens produces images with maximal magnification at its center and severe contraction at its periphery. Nearer images are disproportionally enlarged and remote images are falsely miniaturized.

Foreword

This optical illusion may surprise a surgeon who is not familiar with this optically distorted anatomy at the skull base. It is quite timely that the Napoli group, together with Prof. Manfred Tschabitscher of the Institute of Anatomy of the University of Wien, Austria, would produce an anatomical guide for endoscopic transsphenoidal surgery. Prior to performing endoscopic surgical exploration, surgeons should be well versed with endoscopic anatomy in the sinonasal cavity as well as in the intracranial cavity. I congratulate Cappabianca et al. for their contribution to endoscopic pituitary surgery.

Hae-Dong Jho, M.D. & Ph. D.
Professor of Neurosurgery
University of Pittsburgh, PA, USA

Contents

Introduction	IX
Equipment specification	XI

I. Anatomic preparations — 1

  I.A. Gross anatomy — 2

    I.A.1. Bone preparations — 4

    I.A.2. Nose and paranasal sinuses — 22

  I.B. Endoscopic surgical anatomy — 25

    I.B.1. Nose — 29

    I.B.2. Sphenoidal sinus — 30

    I.B.3. Sella turcica region — 31

    I.B.4. Suprasellar region — 33

    I.B.5. Parasellar region — 43

    I.B.6. Retrosellar region — 47

II. Preoperative management

  II.A. Neuroradiological investigations — 53

    II.A.1. CT — 56

    II.A.2. MRI — 62

  II.B. Operating theatre — 67

    II.B.1. Positioning of the patient — 70

    II.B.2. Equipment — 73

III. Surgical procedure — 79

  III.A. Surgical steps — 80

    III.A.1. Endonasal approach to the sphenoidal sinus ostium — 83

    III.A.2. Enlargement of the sphenoidal sinus ostium — 89

    III.A.3. Preparation of the sphenoid sinus — 91

    III.A.4. Opening of the floor of the sella turcica — 93

    III.A.5. Opening of the dura mater — 97

    III.A.6. Removal of the lesion — 99

    III.A.7. Sella turcica reconstruction — 101

Contents

Appendix: Selected clinical cases   103

    Case 1: Intra-suprasellar macroadenoma   104

    Case 2: Intra-parasellar macroadenoma   109

    Case 3: Solid intra-suprasellar craniopharyngeoma   113

    Case 4: Cystic intra-suprasellar craniopharyngeoma   117

    Case 5: Arachnoid intra-suprasellar cyst   121

    Case 6: Intra-suprasellar RATHKE's cleft cyst   125

References   129

Index   132

Introduction

Endoscopic surgery of the pituitary region is a recent evolution of transsphenoidal surgery, pioneered by the anatomical studies of the Italian Davide Giordano, chief surgeon of the Venice Hospital in the late 1800, and developed in the last century by the great names of modern neurosurgery, starting with Cushing, passing through Dott, Guiot and Hardy and progressing to outstanding contemporary figures, such as Laws and Wilson. Transsphenoidal surgery has become a standardized procedure via the sublabial or the transnasal route thanks to the illumination and the magnification provided by the operating microscope and to the intraoperative control of surgical actions by means of fluoroscopy or neuronavigation. It is routinely employed in the treatment of pituitary lesions, thus allowing safer and less traumatic removal of the tumor, as compared to transcranial surgery via a craniotomy.

The transsphenoidal approach using the microscope and the nasal speculum has provided excellent results in large series. It produces minimal trauma and is favoured by the patients. The subsequent development of endoscopy, with its possibility of a vision "inside the anatomy", has made a closer and wider view of the surgical field possible and resulted in even less operative trauma.

The French neurosurgeon Guiot was the first in 1963 to adopt an endoscope in the course of a traditional transsphenoidal approach, to obtain an overview of the contents of the sella turcica. In the subsequent years more endoscope-assisted microneurosurgery or "pure" endoscopic surgery to this area became more frequent, particularly after the contribution of the otorhinolaryngologists and the widespread use of the endoscope in paranasal sinus surgery. In our opinion, the most relevant innovation is the one-nostril endoscopic endonasal transsphenoidal approach introduced by Dr. Jho, which:
- is really minimally traumatizing for the nasal, paranasal and sella turcica structures;
- permits a wider view inside the anatomy, close to the target, without the restriction imposed by the transsphenoidal retractor used in the microsurgical procedure;

Introduction

- opens the doors towards a more complete management of this area and of the surrounding structures;
- is a prelude to the forthcoming "imaging neurosurgery" of the skull base.

We have studied the anatomy of the region in fresh, non-fixed (but for better identification of the arteries they were injected with red latex) cadaver heads and have practiced the endoscopic approach according to Dr Jho's criteria since 1997 in 120 consecutive patients, always recording the anatomical and surgical images.

The resulting work is proposed in this atlas and reflects our experience; we hope that it may act as a guide for those who wish to undertake this surgical procedure hopefully and make an ameliorative contribution.

We begin with the description of the anatomic structures on cadaver specimens; in the following chapter there is a short presentation of the neuroradiological studies prior to the operation and their rationale; then comes the detailed description of the surgical procedure, beginning with the positioning of the patient and proceeding through six cases – two macroadenomas, two craniopharyngeomas, one intra-suprasellar arachnoid cyst, and one intra-suprasellar RATHKE cleft cyst.

Equipment specification

Endoscopic systems used in the Neurosurgical Department:

Endoscopes

When employed in course of surgery, they are inserted in an outer sleeve connected to an irrigation system to clean the front lens. For surgical purposes, endoscopes manufactured by Karl Storz®, Tuttlingen (Germany) have always been used at the Neurosurgical Department of the "Federico II" University of Naples:

a) 0°, 4 mm outer diameter, 180 mm length (Karl Storz®, Tuttlingen, Germany)
b) 30°, 4 mm outer diameter, 180 mm length (Karl Storz®, Tuttlingen, Germany)
c) 45°, 4 mm outer diameter, 180 mm length (Karl Storz®, Tuttlingen, Germany)
d) 70°, 4 mm outer diameter, 180 mm length (Karl Storz®, Tuttlingen, Germany)
e) 0°, 4 mm outer diameter, 300 mm length (Karl Storz®, Tuttlingen, Germany)
f) 30°, 4 mm outer diameter, 300 mm length (Karl Storz®, Tuttlingen, Germany)
g) 70°, 4 mm outer diameter, 300 mm length (Karl Storz®, Tuttlingen, Germany)

Light Fountain

Aesculap® light source 300
Storz® Xenon 300

Video Recording System

DV CAM - DSR-20MDP: Sony®
S-VHS: Sony®
U-matic – High band: Sony®
U-matic – Low band: Sony®

Video-Camera Equipment

Aesculap® CCD-Camera
Storz® Endovision Telecam SL

High-Speed Microdrill

Storz® multidrive II

Equipment specification

Endoscopic systems used in the Anatomic Institute:

Endoscopes

All endoscopes employed are rigid scopes without a working channel. When used for anatomical studies on the cadaver, no irrigation/suction channel is necessary so that the outer diameter is determined by the light fibers and the lenses.

For anatomical studies at the Institute of Anatomy of the University of Vienna the equipment used throughout the preparation consisted of the following endoscopes:

a) 0°, 4 mm outer diameter, 180 mm length (Aesculap®, Tuttlingen, Germany)

b) 30°, 4 mm outer diameter, 180 mm length (Aesculap®, Tuttlingen, Germany)

c) 0°, 4 mm outer diameter, 180 mm length (Olympus®, Austria)

d) 70°, 4 mm outer diameter, 180 mm length (Olympus®, Austria)

e) 0°, 4 mm outer diameter, 180 mm length (Karl Storz®, Tuttlingen, Germany)

f) 30°, 4 mm outer diameter, 180 mm length (Karl Storz®, Tuttlingen, Germany)

g) 70°, 4 mm outer diameter, 180 mm length (Karl Storz®, Tuttlingen, Germany)

Light Fountain

Olympus®: CLV-S

Storz®: Flashgenerator 600

Video Recording System

Sony®: Digital Still Recorder

Sony®: Videocasette Recorder U-matic SP VO 9600 P

Sony®: Videocasette Recorder Beta Cam SP UVW 1800P

Sony®: Videocasette Recorder S-VHS SVO 9500 MDP

Video-Camera Equipment

Storz®: Endovision Telecam SL

Equipment specification

Camera Equipment

Contax® 167 MT

Kodak® DCS 330

Olympus® M10 with an Olympus® SM-ER 2 adapter

Lasergraphics® Personal LFR Plus

Photographic Films

Kodak Ektachrome® 100 and 400 ASA were used throughout

High-Speed Microdrill

Aesculap®: HILAN

I. Anatomic preparations

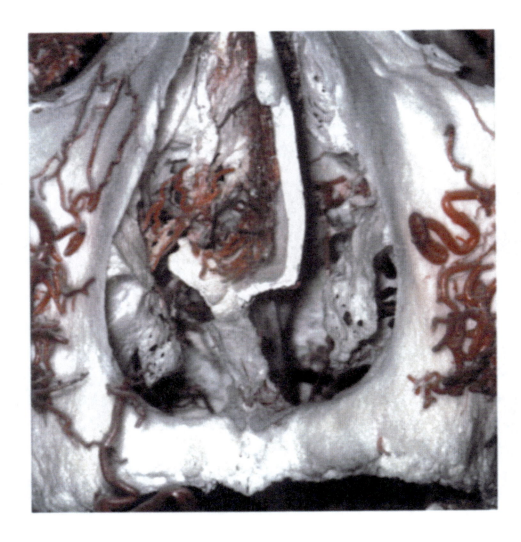

I. Anatomic preparations

I.A. Gross anatomy

The endoscopic endonasal transsphenoidal approach to the pituitary requires the anatomical knowledge not only of its target, for example the sella turcica region. An intimate knowledge of each nasal structure passed through by the surgeon, such as the inner nose and the adjacent paranasal sinuses, is a prerequisite.

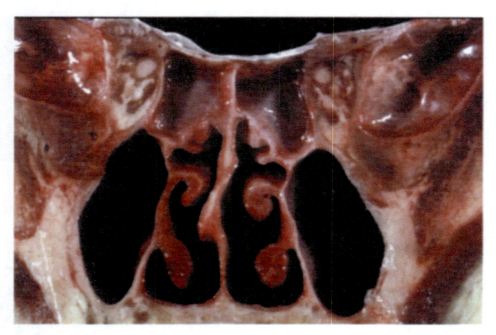

Fig. 1 Coronal section of the nasal and paranasal structures.

I. Anatomic preparations

I.A. Gross anatomy

Introduction

For centuries normal human anatomy, both gross and microscopic, has been studied in depth. Anatomical structures have not changed over time. What has changed is the way of viewing them. The operating microscope opened up an entirely new dimension and new landmarks were needed. The endoscope is posing a similar challenge. Inherent features, like dimension (unlike under the microscope images become larger and brighter the closer the scope comes to them), spatial distorsion (the fish-eye effect blurs the margins) and, last but not least, the various angled designs (0°, 30°, 70° and others) make long-known structures appear in a different perspective.

Simply describing and measuring anatomical details provides a general idea of anatomy, which is reflected in textbooks and atlases of anatomy. But surgeons today have access to imaging techniques, which show them the individual anatomy of a given patient. Together with a profound understanding of anatomy, the combined use of sophisticated imaging techniques (CT, MRI, ultrasound, etc.) is the very basis of successful surgery. What TIEDEMANNS (1781–1861) said is still valid today: "Without anatomy physicians are like moles. They labor in the dark and despite their labors they leave but mounds. (Freely translated from a quotation by KUSSMAUL: Ärzte ohne Anatomie gleichen den Maulwürfen. Sie arbeiten im Dunkel und ihrer Hände Tagwerk sind – Erdhügel).

We are very grateful to Mag. Krista Schmidt and Mrs. Mary Keating for preparing the English-language manuscripts.

Last, but not least, we owe a vote of thanks to all those who, following a long tradition in Vienna, donated their bodies to the Institute of Anatomy for teaching and research. Without them scientific work in anatomy would be unthinkable.

I. Anatomic preparations

I.A. Gross anatomy

I.A.1. Bony preparations

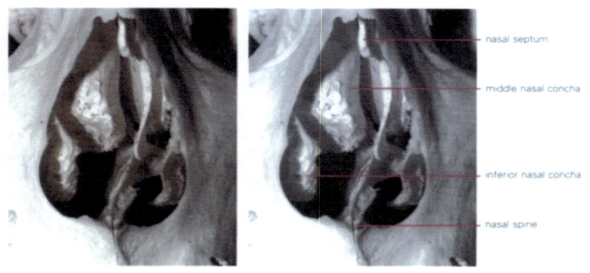

Fig. 2 Piriform recess and nasal bony structures.

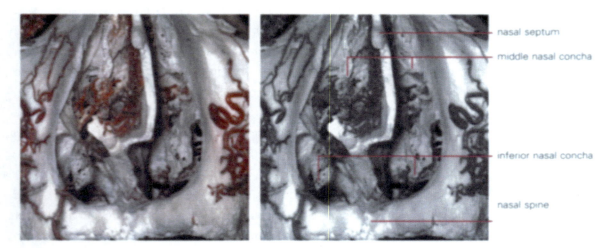

Fig. 3 Note the rich vascularization of the nasal structures.

I. Anatomic preparations
I.A. Gross anatomy
I.A.1. Bony preparations

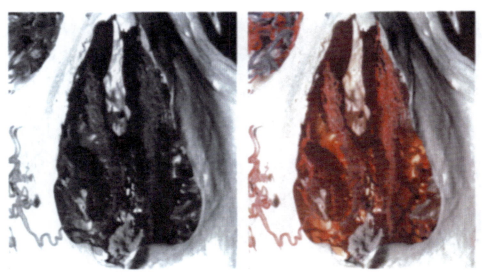

Fig. 4 Note the rich arterial vascularisation of the septal mucosa

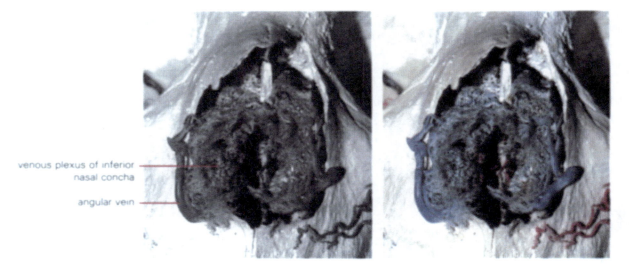

venous plexus of inferior nasal concha

angular vein

Fig. 5 Note the rich submucosal cavernous plexus

I. Anatomic preparations

I.A. Gross anatomy

I.A.1. Bony preparations

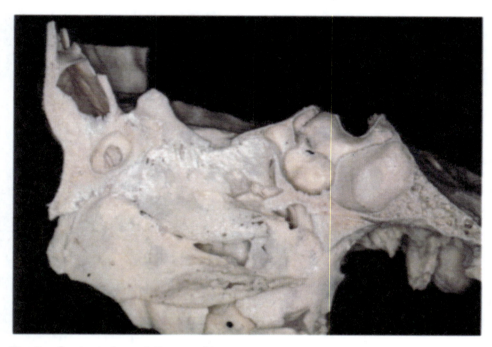

Fig. 6 Sagittal view of the nasal bony structures.

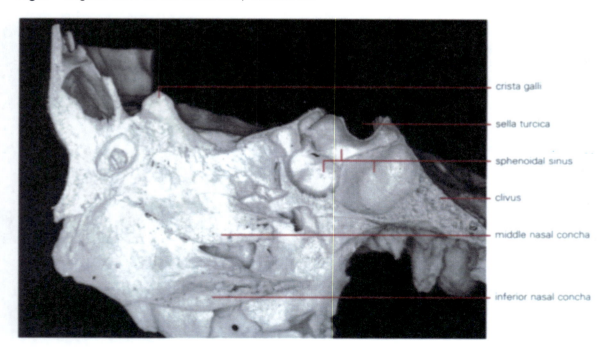

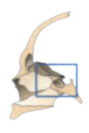

I. Anatomic preparations
I.A. Gross anatomy
I.A.1. Bony preparations

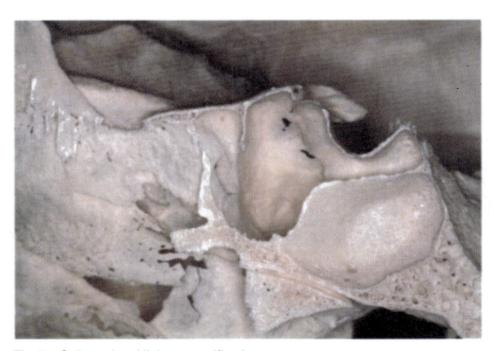

Fig. 7 Sella region. Higher magnification

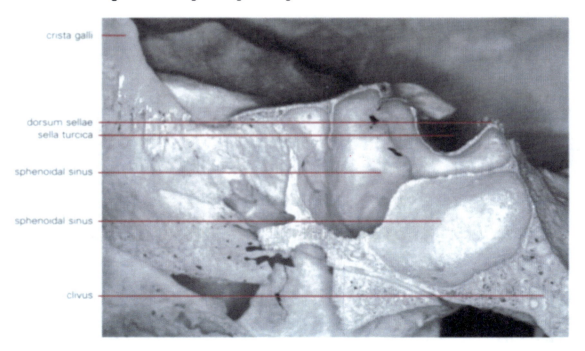

I. Anatomic preparations

I.A. Gross anatomy

I.A.1. Bony preparations

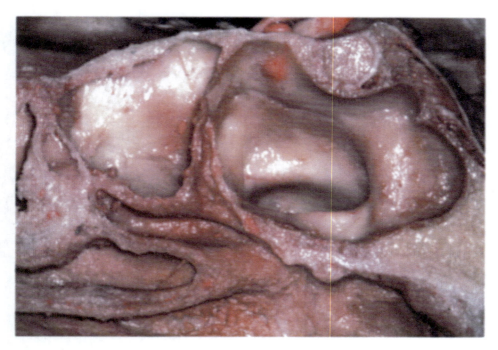

Fig. 8 Sagittal view of the nasal and sphenoidal sinus structures. The mucosa is preserved.

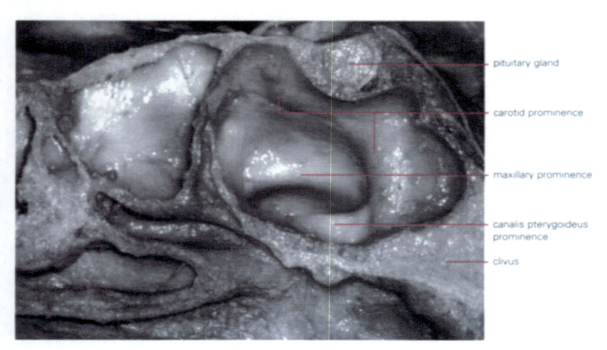

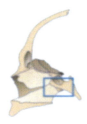

I. Anatomic preparations
I.A. Gross anatomy
I.A.1. Bony preparations

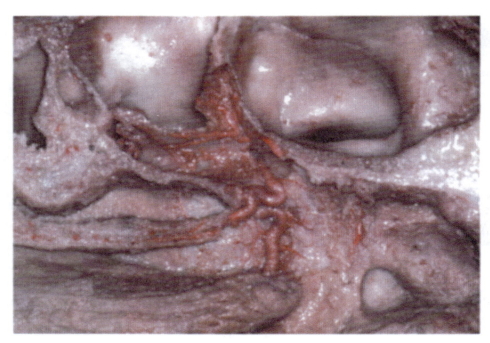

Fig. 9 Sagittal view of the nasal and sphenoid structures. The sphenopalatine artery.

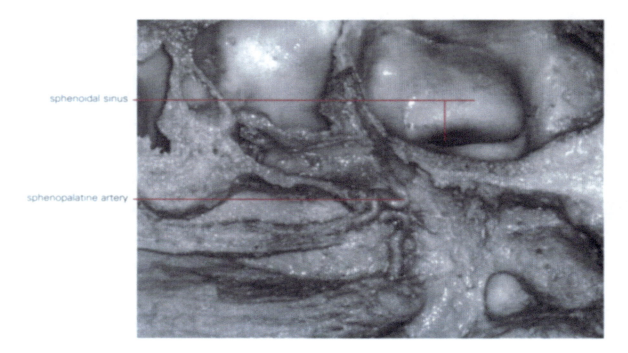

I. Anatomic preparations

I.A. Gross anatomy

I.A.1. Bony preparations

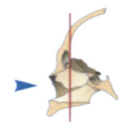

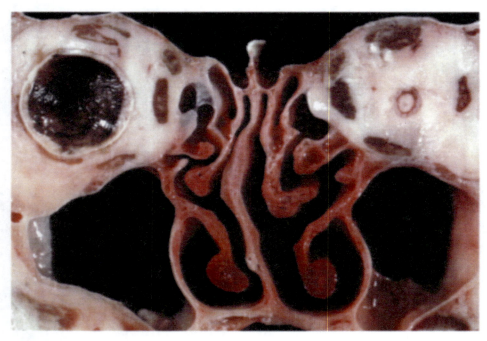

Fig. 10 Coronal section of the nasal and paranasal structures anterior to the sphenoidal sinus.

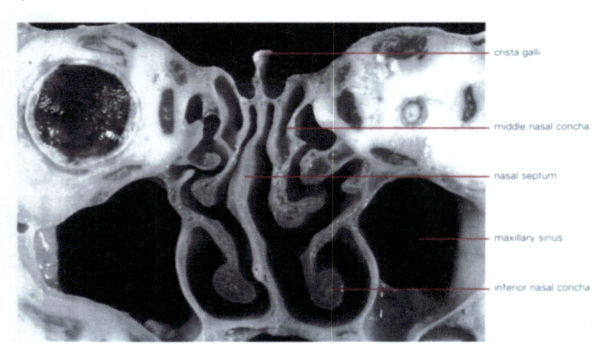

- crista galli
- middle nasal concha
- nasal septum
- maxillary sinus
- inferior nasal concha

I. Anatomic preparations
I.A. Gross anatomy
I.A.1. Bony preparations

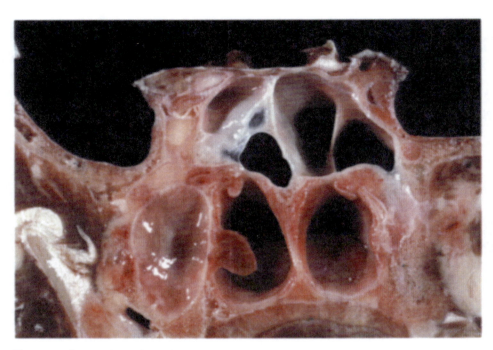

Fig. 11 Coronal section of the sphenoidal sinus anterior to the sella turcica.

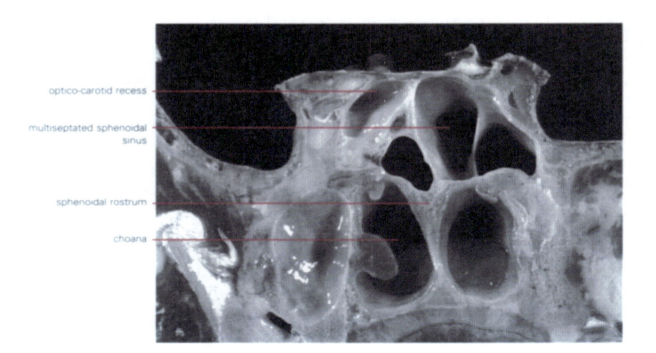

- optico-carotid recess
- multiseptated sphenoidal sinus
- sphenoidal rostrum
- choana

I. Anatomic preparations
I.A. Gross anatomy
I.A.1. Bony preparations

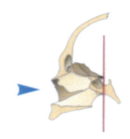

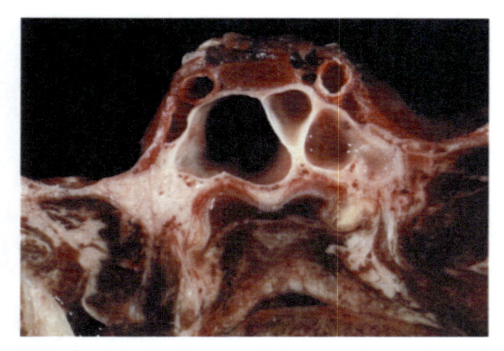

Fig. 12 Coronal section of the sphenoidal sinus centered on the sella turcica.

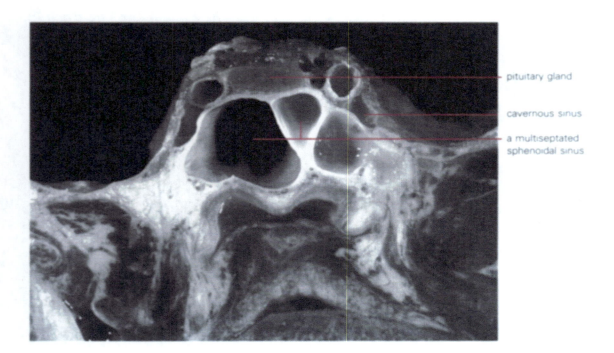

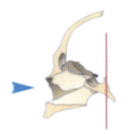

I. Anatomic preparations
I.A. Gross anatomy
I.A.1. Bony preparations

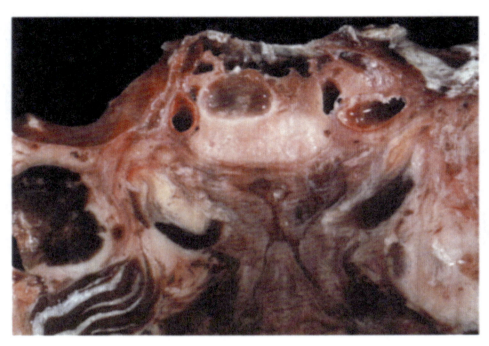

Fig. 13 Coronal section posterior to the sella turcica centered on the clivus.

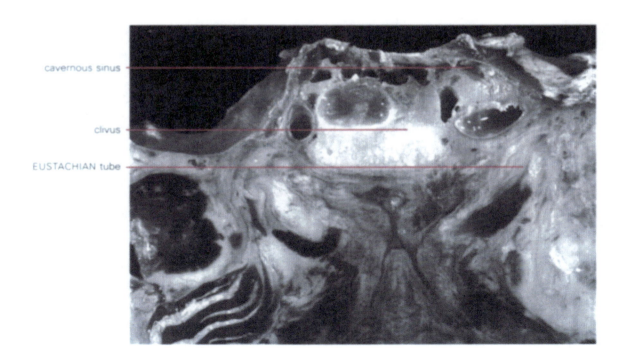

- cavernous sinus
- clivus
- EUSTACHIAN tube

I. Anatomic preparations

I.A. Gross anatomy

I.A.1. Bony preparations

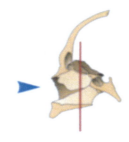

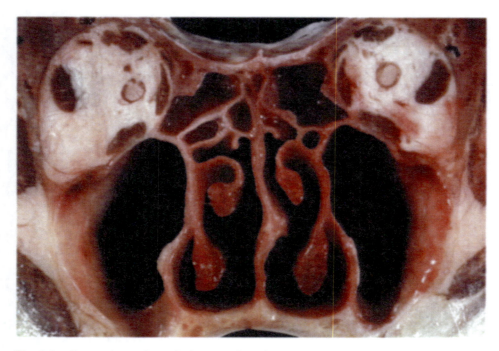

Fig. 14 Coronal section of the nasal and paranasal structures anterior to the sphenoidal sinus.

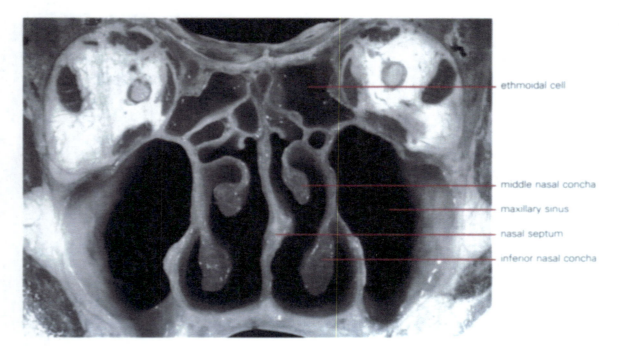

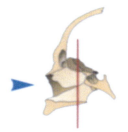

I. Anatomic preparations
I.A. Gross anatomy
I.A.1. Bony preparations

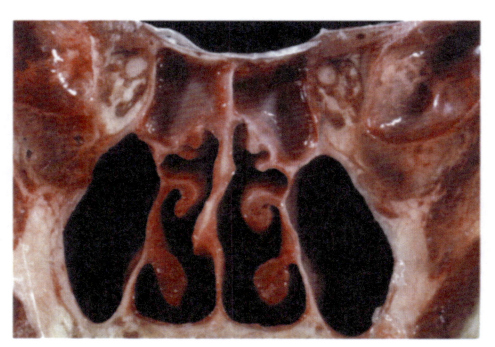

Fig. 15 Coronal section of the nasal and paranasal structures anterior to the sphenoidal sinus.

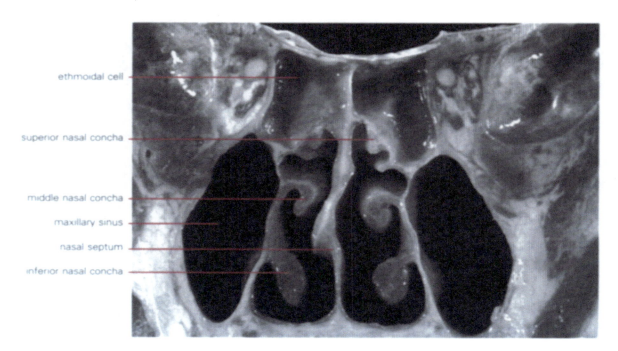

I. Anatomic preparations

I.A. Gross anatomy

I.A.1. Bony preparations

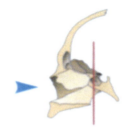

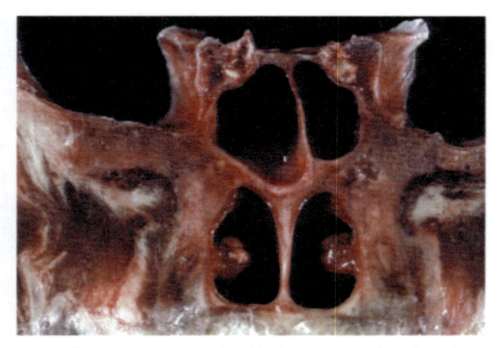

Fig. 16 Coronal section of the sphenoidal sinus anterior to the sella turcica.

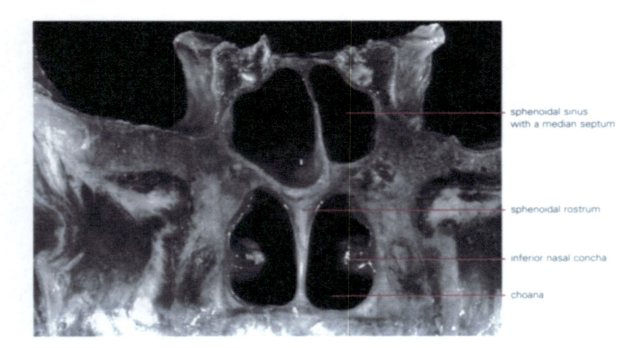

- sphenoidal sinus with a median septum
- sphenoidal rostrum
- inferior nasal concha
- choana

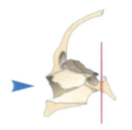

I. Anatomic preparations
I.A. Gross anatomy
I.A.1. Bony preparations

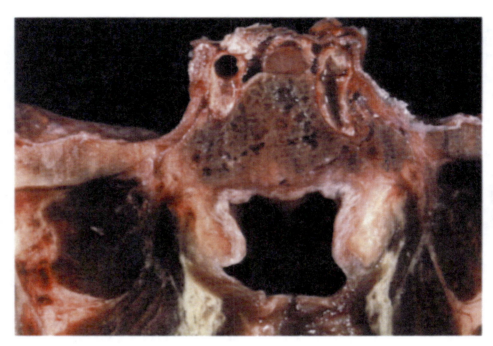

Fig. 17 Coronal section of the posterior part of the sphenoidal sinus which is not pneumatized (conchal type). This section is centred on the sella turcica.

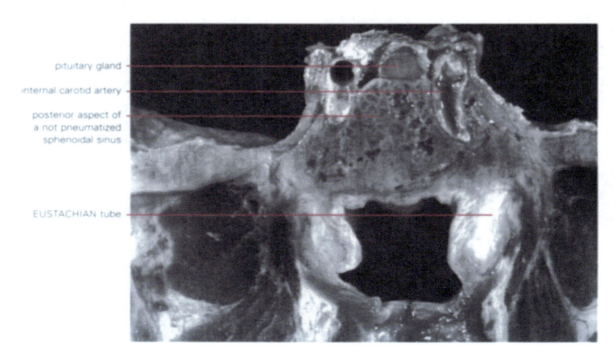

- pituitary gland
- internal carotid artery
- posterior aspect of a not pneumatized sphenoidal sinus
- EUSTACHIAN tube

I. Anatomic preparations

I.A. Gross anatomy

I.A.1. Bony preparations

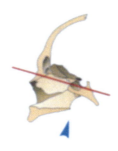

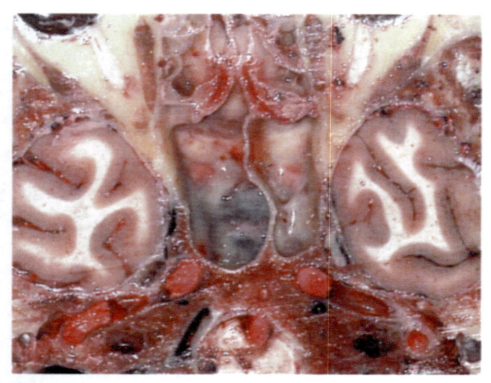

Fig. 18 Transversal section seen from below.

- optic nerve
- carotid artery
- pituitary gland
- carotid canal

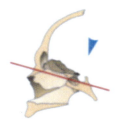

I. Anatomic preparations
I.A. Gross anatomy
I.A.1. Bony preparations

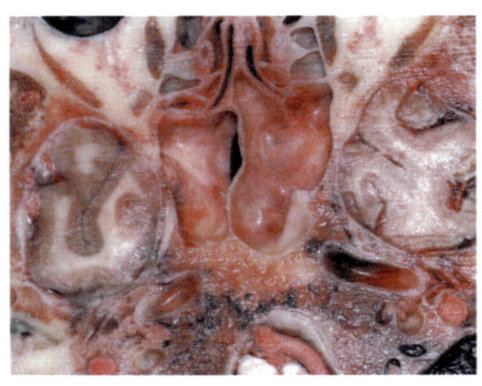

Fig. 19 Transversal section seen from above.

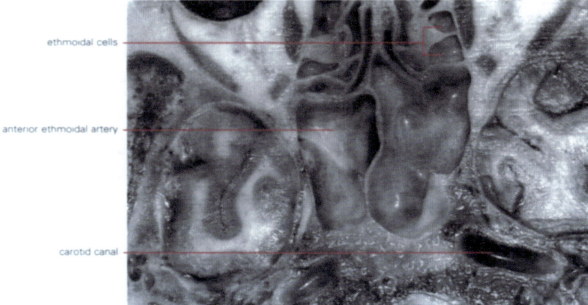

I. Anatomic preparations
I.A. Gross anatomy
I.A.1. Bony preparations

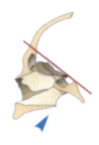

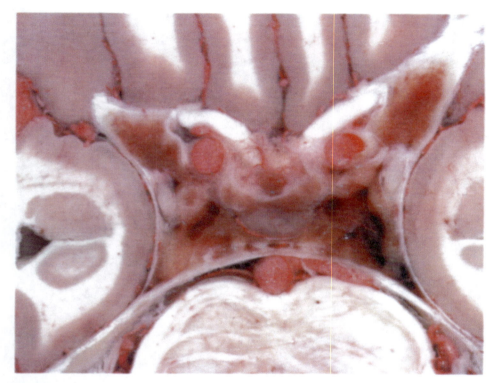

Fig. 20 Transversal section seen from below.

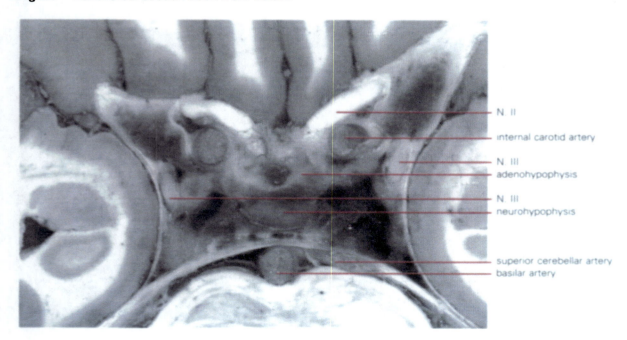

- N. II
- internal carotid artery
- N. III
- adenohypophysis
- N. III
- neurohypophysis
- superior cerebellar artery
- basilar artery

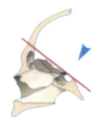

I. Anatomic preparations
I.A. Gross anatomy
I.A.1. Bony preparations

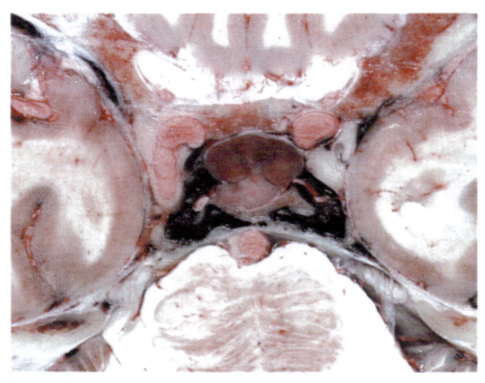

Fig. 21 Transversal section seen from above.

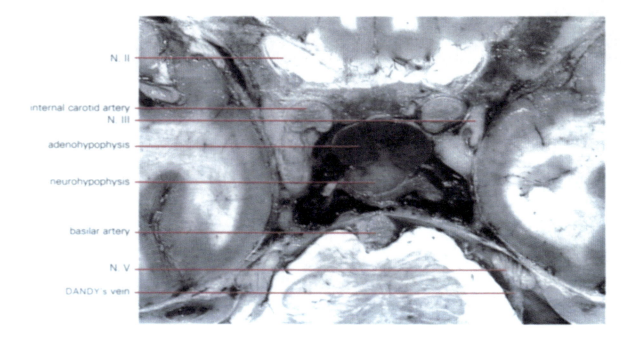

I. Anatomic preparations

I.A. Gross anatomy

I.A.2. Nose and paranasal sinuses

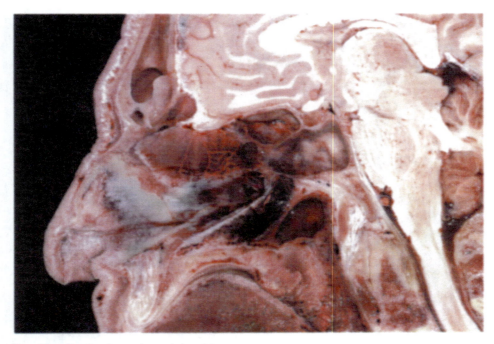

Fig. 22 Sagittal section, right side.

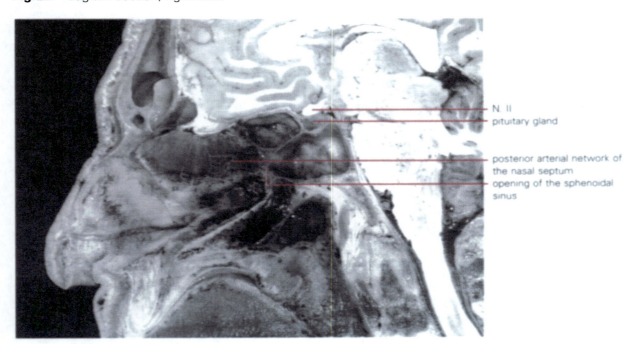

I. Anatomic preparations
I.A. Gross anatomy
I.A.2. Nose and paranasal sinuses

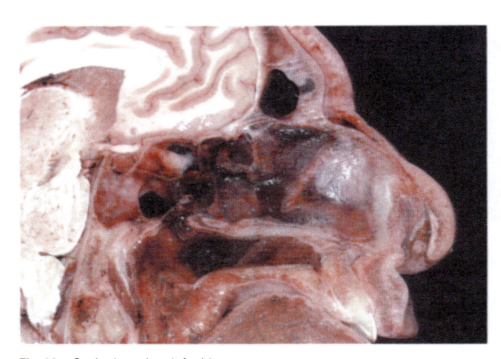

Fig. 23 Sagittal section, left side.

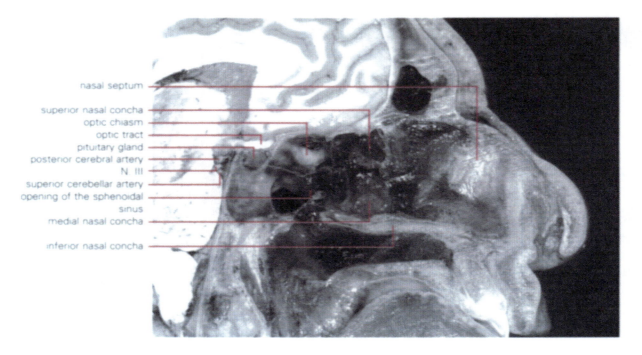

- nasal septum
- superior nasal concha
- optic chiasm
- optic tract
- pituitary gland
- posterior cerebral artery
- N. III
- superior cerebellar artery
- opening of the sphenoidal sinus
- medial nasal concha
- inferior nasal concha

23

I. Anatomic preparations

I.B. Endoscopic surgical anatomy

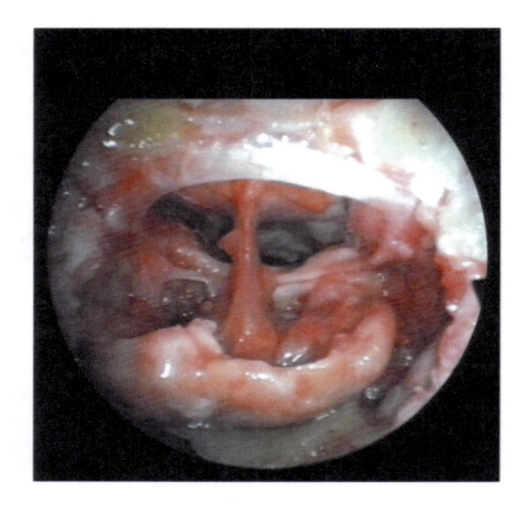

I. Anatomic preparations

I.B. Endoscopic surgical anatomy

Nose

After introduction of the endoscope into the nostril the structures met are the septum medially and the inferior nasal concha laterally. The first important structure to identify is the middle nasal concha, which represents a landmark in the determination of the direction of the endoscope. The space between the septum and the middle nasal concha is usually very narrow. Its lateral luxation may be needed to create space to push the endoscope forward. By following the inferior margin of the middle nasal concha the upper edge of the nasal choana is identified. When the spheno-ethmoidal recess appears, it is sufficient to alter the course of the endoscope in a caudia-cranial (inferior-superior) direction to visualize the sphenoidal sinus ostium. The sphenoidal sinus ostium appears on each side in the upper part of the anterior wall of the sphenoidal sinus and may be hidden by the superior nasal concha. Sometimes the sphenoidal sinus ostium cannot be identified. In this case an anterior sphenoidectomy has to be done in the upper part of the choana, and this, with the inferior margin of the middle nasal concha, is our constant landmark.

Sphenoidal sinus

The natural ostium of the sphenoid is enlarged using KERRISON's rongeurs and the sphenoidal rostrum is removed. When the anterior wall of the sphenoidal sinus is removed, the sphenoidal sinus is exposed. The sphenoidal sinus may be well pneumatized (sellar type), partially pneumatized (presellar type), or not pneumatized at all (conchal type). Usually one or more septa subdivide the sphenoidal sinus, but sometimes no septa are found. The septum of the sphenoidal sinus is removed and the posterior wall of the sphenoidal sinus is exposed. The posterior aspect of the planum sphenoidale is localized anteriorly in the roof, the sella turcica in the middle and the clivus posteriorly and caudally. At the posterolateral walls of the sphenoidal sinus bilaterally the optic protuberances, the optico-carotid recesses and the carotid protuberances cranio-caudally can be identified.

I. Anatomic preparations

I.B. Endoscopic surgical anatomy

Sella turcica region

The sella turcica is opened using a high-speed micro-drill and the micro-KERRISON rongeurs or other punches such as the STAMMBERGER circular cutting punch®. The dura mater of the floor of the sella turcica is incised and the pituitary gland is seen, yellowish in color and of soft consistency. Displacing the pituitary gland medially and upwards, the inferior hypophyseal arteries, originating from the meningohypophyseal trunk or directly from the intracavernous carotid artery, are observed.

Supra-, para- and retro-sellar region

To reach the suprasellar cistern in the suprasellar area the tuberculum must be drilled from below to expose the diaphragma sellae. With the separation of the diaphragma from the gland the suprasellar cistern appears. After opening the cistern the pituitary stalk with its peculiar aspect is disclosed together with the superior hypophyseal arteries. On further inspection the whole suprasellar region can be seen. Above the pituitary stalk the optic chiasm is identified. The further introduction of the endoscope up to the optic chiasm exposes the anterior circle of Willis and the gyrus rectus bilaterally.

The removal of the bone of the lateral wall of the sphenoidal sinus exposes the parasellar area. Rostro-caudally in the parasellar area the optic nerve, the ophthalmic artery and the intracavernous carotid artery can be distinguished.

Going further in a lateral direction and displacing the carotid artery medially the nerves in the wall and in the cavernous sinus are identified. In a rostro-caudal direction, the oculomotor, the trochlear, the first branch of the trigeminal, and the abducent nerves are identified. All these nerves of the cavernous sinus converge towards the superior orbital fissure.

I. Anatomic preparations

I.B. Endoscopic surgical anatomy

Finally the 0° and 30° endoscopes are introduced into the retrosellar space and the basilar tip and the posterior circle of Willis are visualized. The floor of the third ventricle is opened and the ventricular cavity explored.

I. Anatomic preparations
I.B. Endoscopic surgical anatomy
I.B.1. Nose

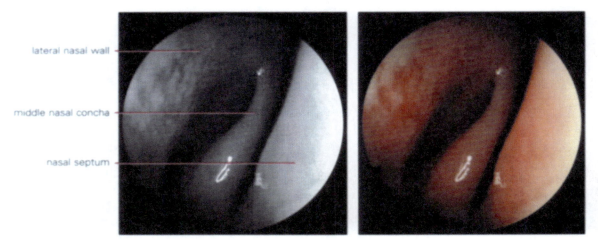

Fig. 24 Right nostril, 0° endoscope. The middle nasal concha is the first landmark to be identified. By following its inferior margin the spheno-ethmoidal recess is reached.

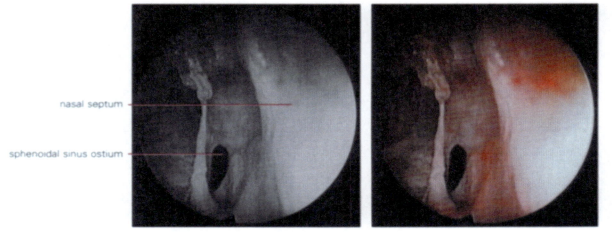

Fig. 25 Right nostril, 0° endoscope. The sphenoidal sinus ostium located at the upper part of the spheno-ethmoidal recess is the crucial point where the procedure starts.

I. Anatomic preparations

I.B. Endoscopic surgical anatomy

I.B.2. Sphenoidal sinus

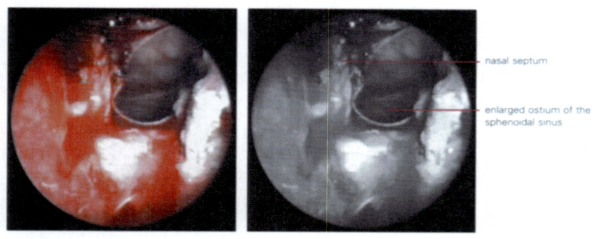

Fig. 26 Left nostril, 0° endoscope. The sphenoidal sinus ostium is broadened using KERRISON's rongeurs.

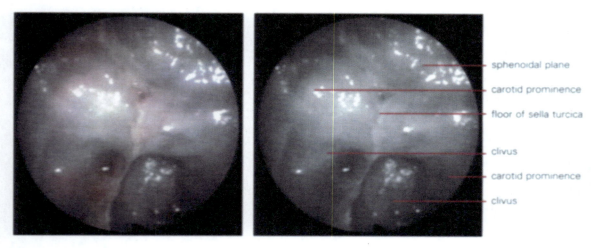

Fig. 27 0° endoscope. Posterior wall of the sphenoidal sinus after removal of the septum of the sphenoidal sinus.

I. Anatomic preparations
I.B. Endoscopic surgical anatomy
I.B.3. Sella turcica region

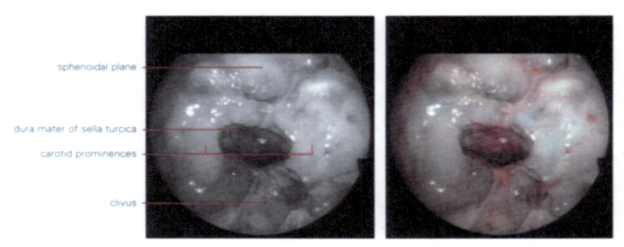

Fig. 28 0° endoscope. The floor of the sella turcica has been removed.

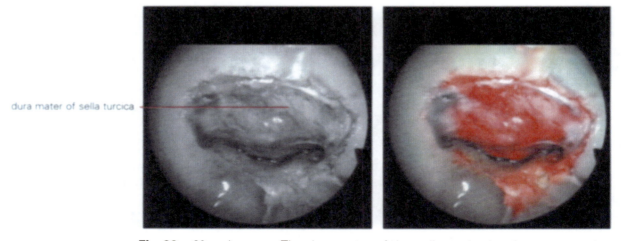

Fig. 29 0° endoscope. The dura mater of the sella turcica has been exposed.

I. Anatomic preparations

I.B. Endoscopic surgical anatomy

I.B.3. Sella turcica region

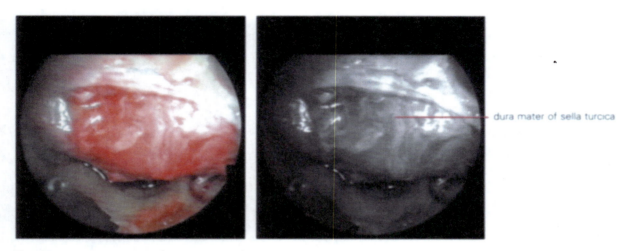

Fig. 30 0° endoscope. Closer view of the dura mater.

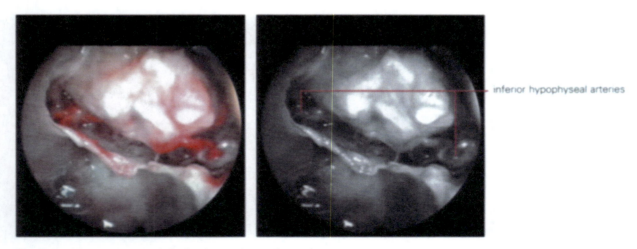

Fig. 31 0° endoscope. Inferior hypophyseal arteries.

I. Anatomic preparations
I.B. Endoscopic surgical anatomy
I.B.4. Surpasellar region

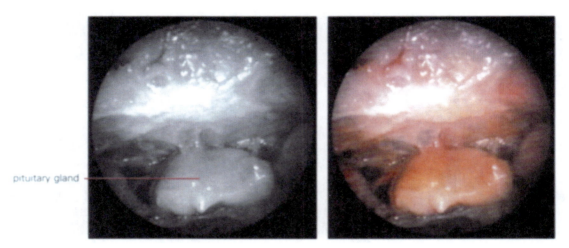

Fig. 32 0° endoscope. The pituitary gland after dura mater removal.

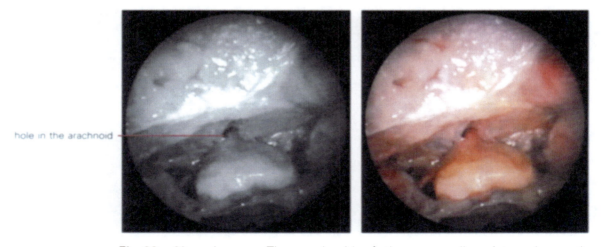

Fig. 33 0° endoscope. The arachnoid of the suprasellar cistern is gently opened.

I. Anatomic preparations

I.B. Endoscopic surgical anatomy

I.B.4. Suprasellar region

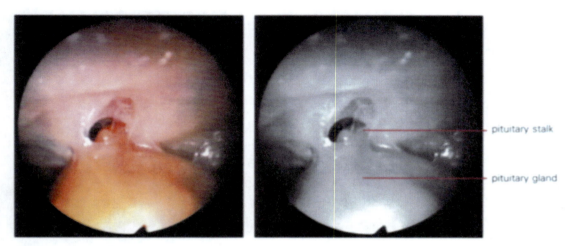

Fig. 34 0° endoscope. A closer approach of the endoscope allows visualization of the pituitary stalk.

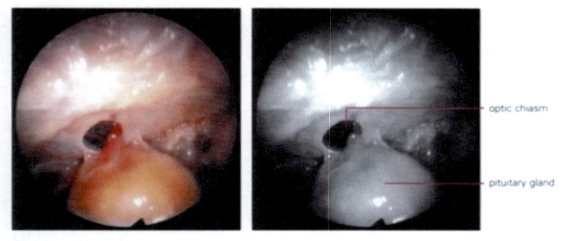

Fig. 35 0° endoscope. Widening of the hole in the arachnoid. The optic chiasm appears.

I. Anatomic preparations
I.B. Endoscopic surgical anatomy
I.B.4. Suprasellar region

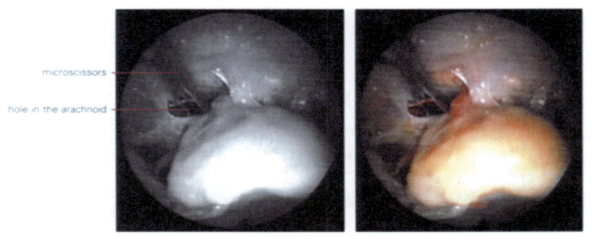

Fig. 36   0° endoscope. Microscissors are used to cut the arachnoid.

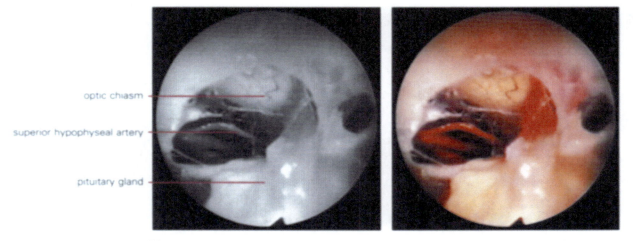

Fig. 37   0° endoscope. The right superior hypophyseal artery is identified.

I. Anatomic preparations

I.B. Endoscopic surgical anatomy

I.B.4. Suprasellar region

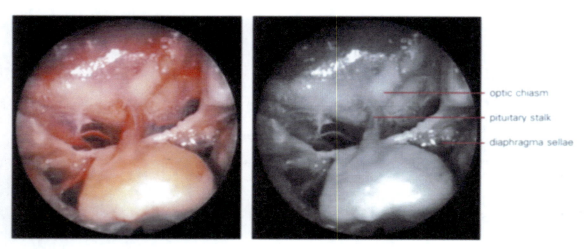

Fig. 38 0° endoscope. The optic chiasm and the pituitary stalk appear after the arachnoid dissection.

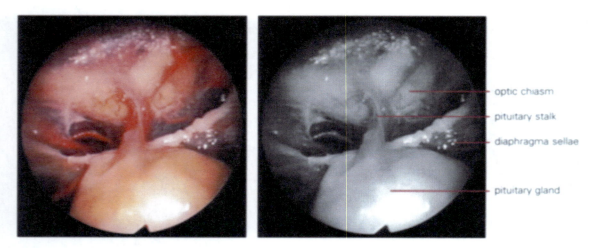

Fig. 39 0° endoscope. Closer view of the opto-chiasmatic region.

I. Anatomic preparations

I.B. Endoscopic surgical anatomy

I.B.4. Suprasellar region

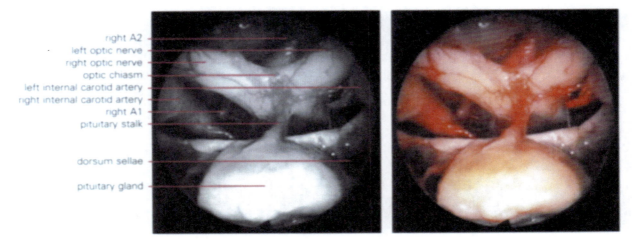

Fig. 40 0° endoscope. The chiasmatic cistern has been opened. Closer view of the pituitary gland and the opto-chiasmatic region. A2 = anterior cerebral artery.

I. Anatomic preparations

I.B. Endoscopic surgical anatomy

I.B.4. Suprasellar region

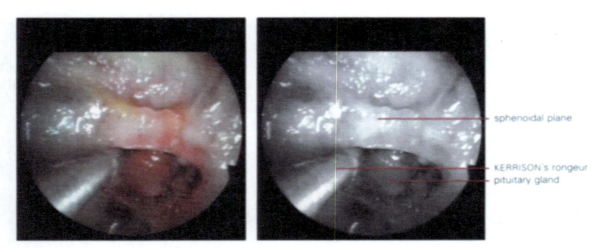

Fig. 41 0° endoscope. Opening of the sphenoidal plane using KERRISON's rongeurs.

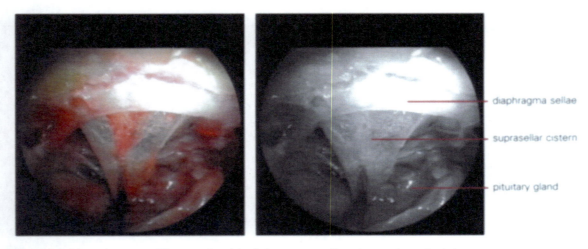

Fig. 42 0° endoscope. The arachnoid of the suprasellar cistern is pulled down.

I. Anatomic preparations

I.B. Endoscopic surgical anatomy

I.B.4. Suprasellar region

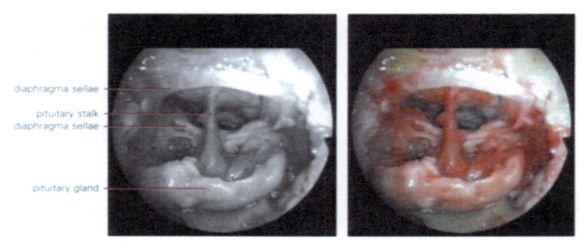

Fig. 43 0° endoscope. The arachnoid has been opened and the pituitary stalk appears.

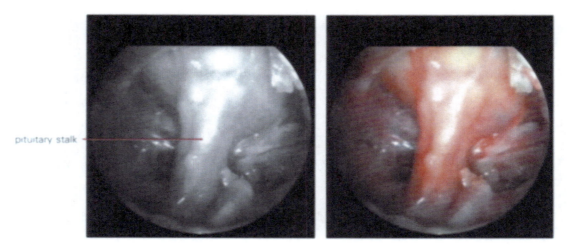

Fig. 44 30° endoscope orientated upwards. Closer view of the pituitary stalk.

I. Anatomic preparations

I.B. Endoscopic surgical anatomy

I.B.4. Suprasellar region

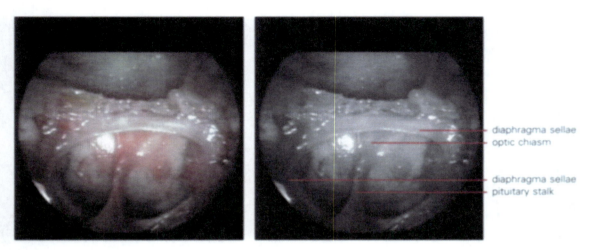

Fig. 45 70° endoscope. The introduction of the 70° endoscope orientated upwards allows a view of the optic chiasm.

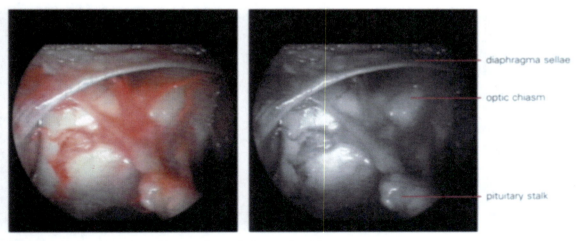

Fig. 46 70° endoscope. Closer view of the optic chiasm.

I. Anatomic preparations
I.B. Endoscopic surgical anatomy
I.B.4. Suprasellar region

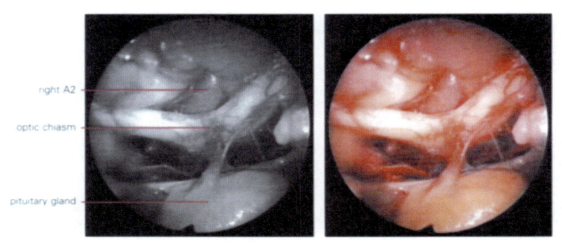

Fig. 47 0° endoscope. The endoscope is directed towards the optic chiasm to expose the anterior circle of Willis. A2 = anterior cerebral artery

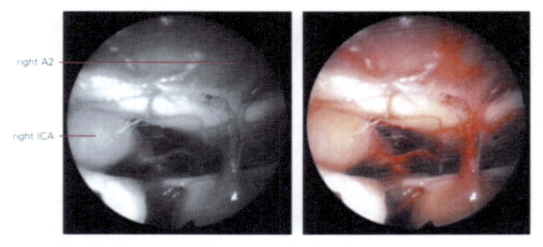

Fig. 48 0° endoscope. Further approximation of the endoscope to the anterior circle of Willis. A2 = anterior cerebral artery, ICA = internal carotid artery.

I. Anatomic preparations

I.B. Endoscopic surgical anatomy

I.B.4. Suprasellar region

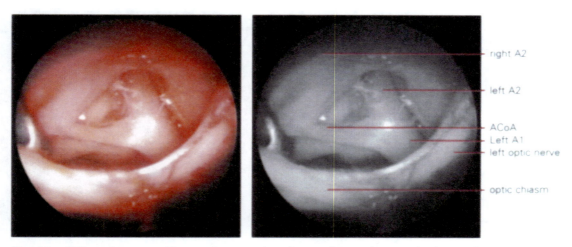

Fig. 49 30° endoscope. The endoscope has been directed towards the optic chiasm. With this angled lens orientated upwards it is possible to visualize the anterior communicating artery (ACoA) and the segment A2 of the anterior cerebral arteries. An instrument has been introduced to dislocate the right A2, which covers the left A2.

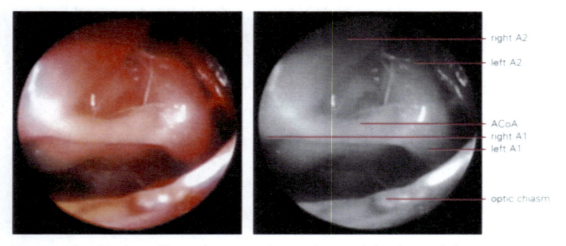

Fig. 50 30° endoscope. The endoscope, orientated upwards is advanced closer to the anterior circle of Willis. A1-2 = anterior cerebral artery, ACoA = anterior communicating artery.

I. Anatomic preparations
I.B. Endoscopic surgical anatomy
I.B.5. Parasellar region

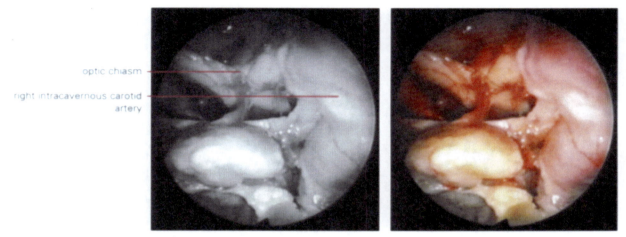

Fig. 51 0° endoscope. The bone of the carotid protuberances has been removed.

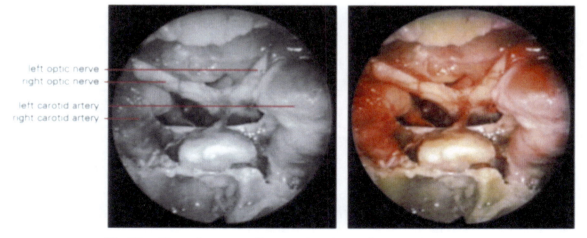

Fig. 52 0° endoscope. Panoramic view of the sella turcica and parasellar region after bone removal.

I. Anatomic preparations

I.B. Endoscopic surgical anatomy

I.B.5. Parasellar region

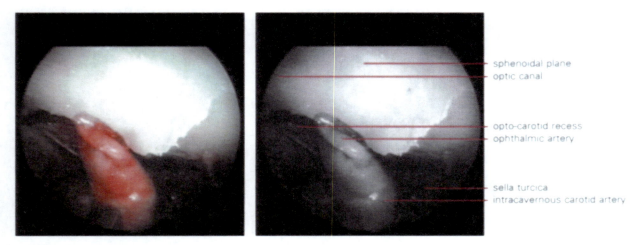

Fig. 53 0° endoscope. Right side. Removal of the upper bony carotid protuberance with view of the right ophthalmic artery.

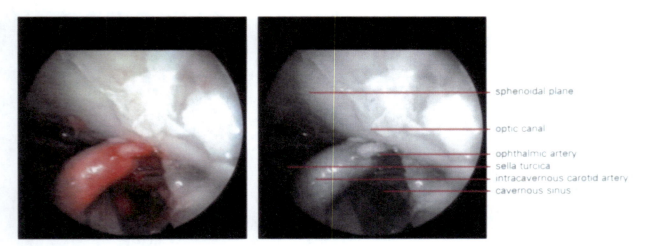

Fig. 54 0° endoscope. Left side. After the removal of the upper bony carotid protuberance the left intracavernous carotid artery has been displaced downward and medially, in order to expose the ophthalmic artery.

I. Anatomic preparations
I.B. Endoscopic surgical anatomy
I.B.5. Parasellar region

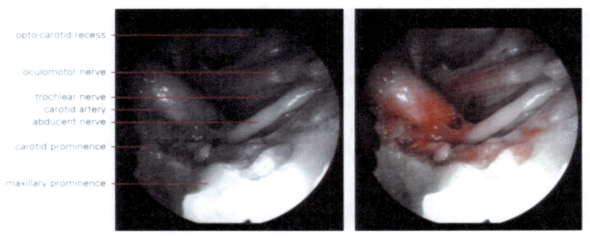

Fig. 55 0° endoscope. Left side. The intracavernous carotid artery has been displaced medially towards the sella turcica to visualize the intracavernous nerves.

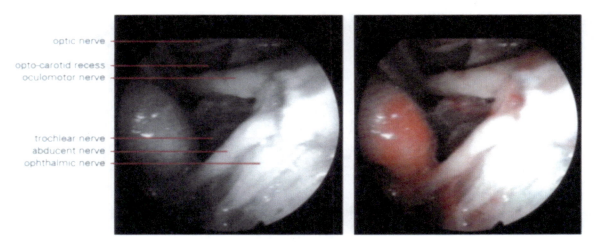

Fig. 56 0° endoscope. Left side. Closer view of the cavernous sinus after medial displacement of the intracavernous carotid artery. The nerves converge towards the medial aspect of the superior orbital fissure.

I. Anatomic preparations

I.B. Endoscopic surgical anatomy

I.B.5. Parasellar region

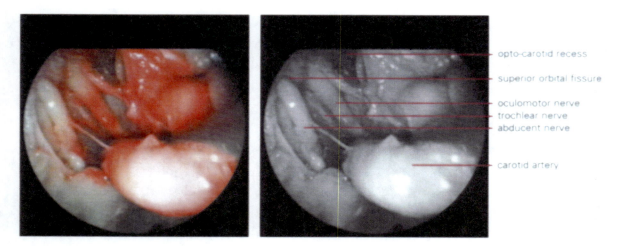

Fig. 57 0° endoscope. Right side. The intracavernous carotid artery has been displaced medially, towards the sella turcica, to expose the intracavernous nerves.

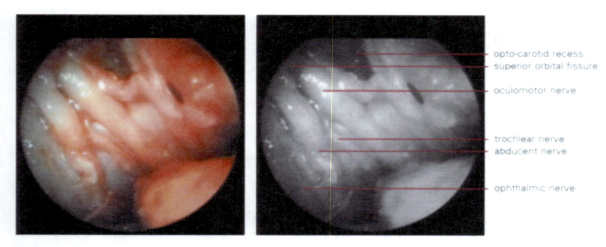

Fig. 58 30° endoscope turned to the right side. The intracavernous carotid artery has been displaced medially towards the sella turcica.

46

I. Anatomic preparations
I.B. Endoscopic surgical anatomy
I.B.6. Retrosellar region

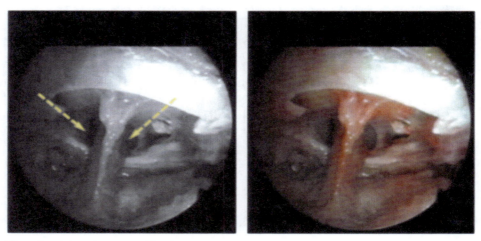

Fig. 59 0° endoscope. The interpeduncular cistern has been opened. The endoscope is introduced behind the pituitary stalk up to the dorsum sellae on both sides (arrows).

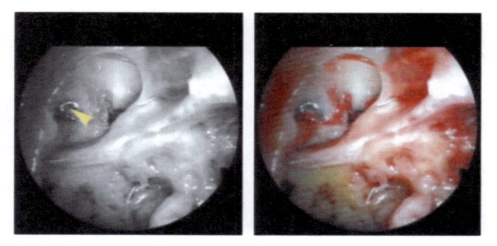

Fig. 60 0° endoscope. The floor of the third ventricle is identified (arrow).

I. Anatomic preparations

I.B. Endoscopic surgical anatomy

I.B.6. Retrosellar region

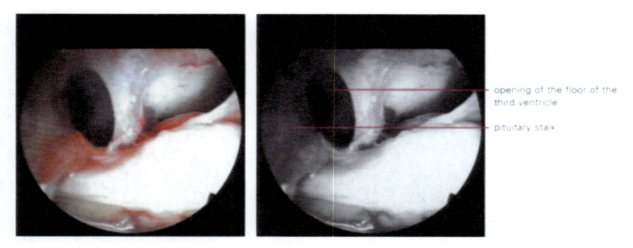

Fig. 61 0° endoscope. The floor of the third ventricle has been opened behind the pituitary stalk.

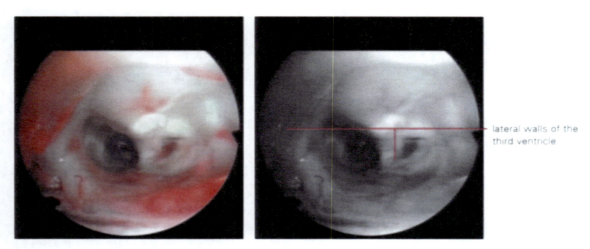

Fig. 62 0° endoscope. The endoscope is inserted into the third ventricle.

I. Anatomic preparations

I.B. Endoscopic surgical anatomy

I.B.6. Retrosellar region

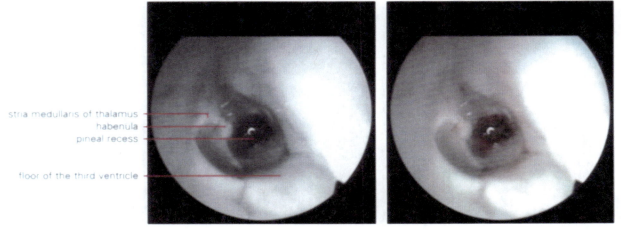

Fig. 63 0° endoscope. The endoscope is inserted into the third ventricle behind the pituitary stalk and from the left side. The posterior part of the ventricular cavity is explored.

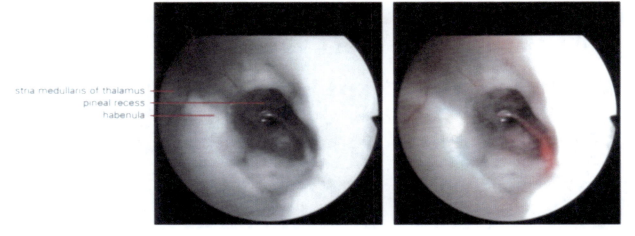

Fig. 64 0° endoscope. The endoscope is approached closer to the pineal recess.

I. Anatomic preparations

I.B. Endoscopic surgical anatomy

I.B.6. Retrosellar region

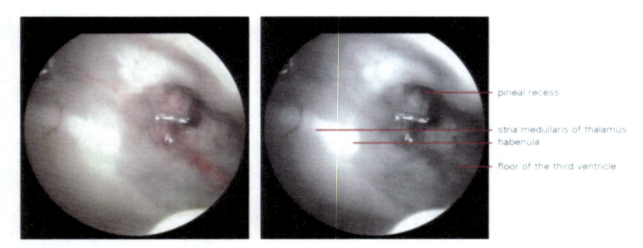

Fig. 65 30° endoscope. The endoscope is turned to the right side.

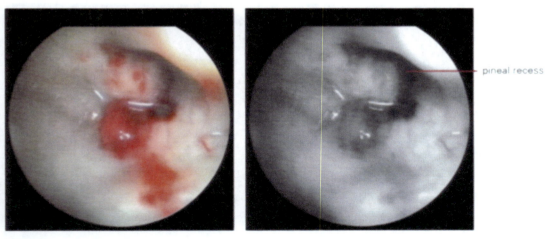

Fig. 66 30° endoscope. The endoscope is orientated upwards and approached closer to the pineal recess. Closer view.

I. Anatomic preparations

I.B. Endoscopic surgical anatomy

I.B.6. Retrosellar region

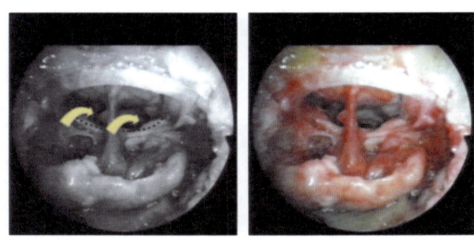

Fig. 67 0° endoscope. In the next pictures the retrosellar area is presented when an endoscope is inserted behind the pituitary stalk and orientated downwards (arrows). The dorsum sellae is outlined with a dotted line.

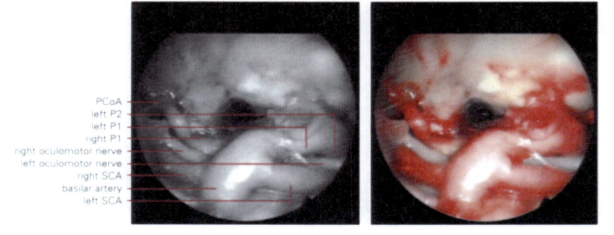

Fig. 68 30° endoscope. After the introduction of a downward orientated endoscope behind the dorsum sellae the basilar tip is visualized. PCoA = posterior communicating artery, SCA = superior cerebellar artery, P1-P2 = posterior cerebral artery.

I. Anatomic preparations

I.B. Endoscopic surgical anatomy

I.B.6. Retrosellar region

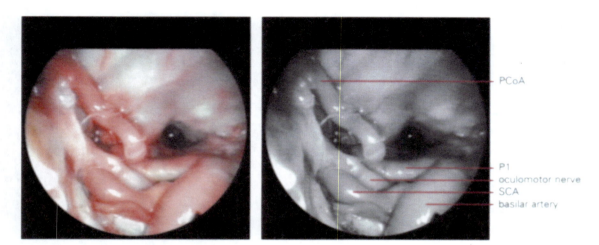

Fig. 69 30° endoscope. Right side, closer view. PCoA = posterior communicating artery, SCA = superior cerebellar artery, P1 = posterior cerebral artery.

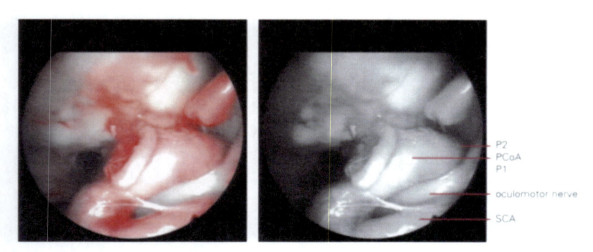

Fig. 70 30° endoscope. Left side, closer view. SCA = superior cerebellar artery, P1-P2 = posterior cerebral artery, PCoA = posterior communicating artery.

II. Preoperative management

II.A. Neuroradiological investigations

The endoscopic endonasal transsphenoidal approach to the pituitary requires the anatomical knowledge not only of its target, for example the sella turcica region. An exact knowledge of all the nasal structures the surgeon will pass through, such as the inner nose and the adjacent paranasal sinuses, is demanded.

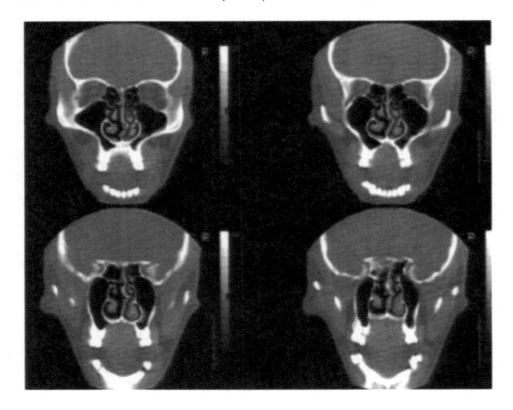

II. Preoperative management

II.A. Neuroradiological investigations

The exact knowledge of the anatomy of the nose, the paranasal sinuses and the sella turcica region of each patient is essential for the correct execution and a good result of the operation.

The anatomical preoperative study must include a **CT scan of nose and paranasal sinuses** with thin slices in coronal and axial sections. Particular attention must be paid to the nasal septum, the nasal conchae, the sphenoidal sinus and the posterior ethmoidal cells. The nasal septum may be deviated or widened, or present a septal spur. When the nasal conchae, particularly the middle nasal concha, are hypertrophic or pneumatized (concha bullosa), the nasal cavity is particularly narrow, and a partial middle turbinectomy may be necessary. The sphenoidal sinus may be well pneumatized (sellar type) or partially pneumatized (presellar type) or, rarely, not pneumatized at all (conchal type), and it is subdivided by one or more septa. Knowledge of the presence and the course of the septum of the sphenoidal sinus or septa gives accurate orientation during surgery and prevents dangerous manoeuvres such as penetration of the septa on the carotid protuberances.

On rare occasions the carotid canal is not ossified and this must be recognized and borne in mind while dissecting the mucosa of the sphenoidal sinus during surgery.

The posterior ethmoidal cells are located anteriorly and lateral to the sphenoidal sinus. The posterior ethmoidal cells may extend further in a posterior direction, superiorly or laterally to the sphenoidal sinus – close to the optic canal as ONODI cells. The presence of polyps in the inner nose or into the paranasal sinuses necessitates their removal. There is adequate space in both these areas to perform the operation and avoid healing problems and sinusitis.

II. Preoperative management

II.A. Neuroradiological investigations

MRI of the sella turcica area, before and after of intravenous paramagnetic contrast enhancement, gives complete depiction of the sella turcica, supra and parasellar structures and their mutual relationships. The pituitary gland, the pituitary stalk and their eventual displacement are well seen on MRI, as they usually present a more intense post-contrast enhancement.

The MRI shows the presence of dislocation of the optic chiasm and the optic nerves and the involvement of the suprasellar structures and third ventricle. When a parasellar extension is present, the displacement or involvement of the ICA and the nerves is depicted.

It is necessary to pinpoint both the parasellar growth of the lesion and the nasal anatomy, as the choice of which nostril to use for the approach may be important in the surgical planning. In fact, since the procedure is not strictly median, and because of the presence of the septum of the sphenoidal sinus, the movements of the endoscope and of the instruments towards the ipsilateral parasella turcica area can be limited, if a wider exposure is not performed. For this reason, when the MRI shows a right parasellar extension, we will use the left nostril and vice versa. When there is no parasellar extension, we choose the wider nostril according to the CT study.

II. Preoperative management

II.A. Neuroradiological investigations

II.A.1 CT

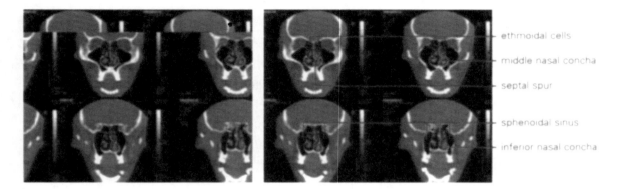

Fig. 71 Coronal serial CT slices of the maxillofacial structures. This study is mandatory to plan the endonasal approach.

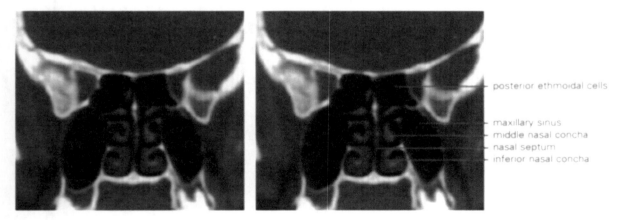

Fig. 72 Coronal CT slice anterior to the sphenoidal sinus. The nasal septum is without deviation.

II. Preoperative management
II.A. Neuroradiological investigations
II.A.1 CT

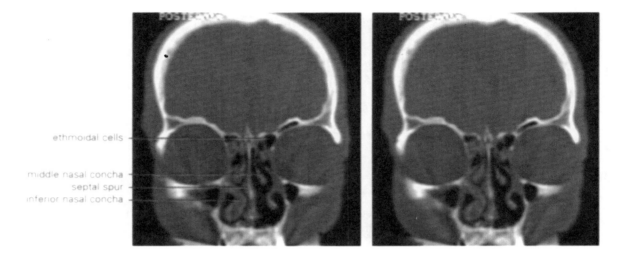

Fig. 73 Maxillofacial CT scan, coronal slice, anterior to the sphenoidal sinus. The right nostril is very narrow because of a large inferior nasal concha and a septal spur.

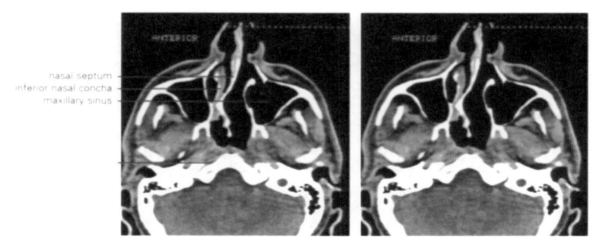

Fig. 74 Maxillofacial CT scan, axial slice. Strong deviation of the nasal septum to the left side.

II. Preoperative management

II.A. Neuroradiological investigations

II.A.1 CT

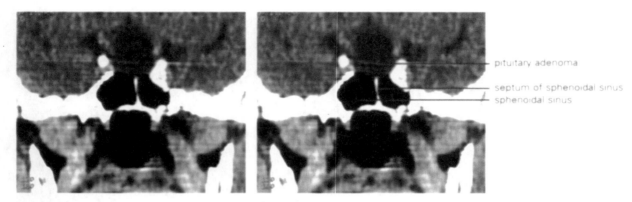

Fig. 75 CT scan. Axial slice centered on the sphenoidal sinus. A median septum is inserted in the midline and the floor of the sella turcica is eroded.

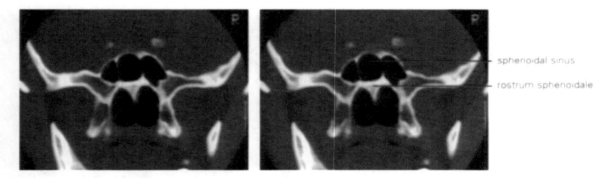

Fig. 76 CT scan (bone window). Coronal slice centered on the sphenoidal sinus. Two lateral sphenoid septa can be observed (=multiseptated).

II. Preoperative management
II.A. Neuroradiological investigations
II.A.1 CT

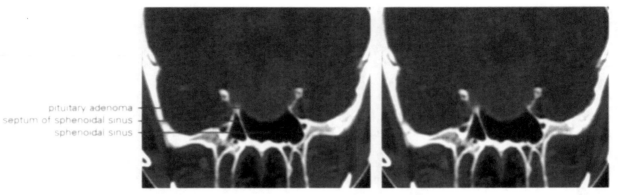

Fig. 77 CT scan. Coronal slice. There is only one septum in the sphenoidal sinus inserted very laterally on the right side. The floor of sella turcica is completely destroyed.

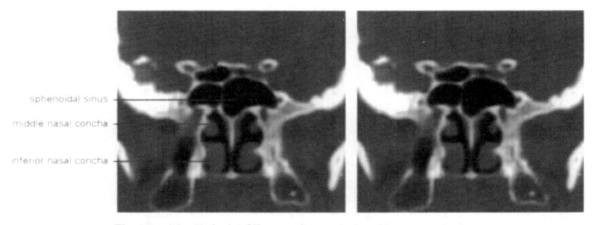

Fig. 78 Maxillofacial CT scan. Coronal slice. The arrow indicates an ONODI cell.

II. Preoperative management

II.A. Neuroradiological investigations

II.A.1 CT

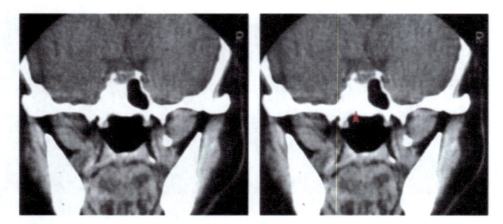

Fig. 79 CT scan. Coronal slice. Part of the sphenoidal sinus is not pneumatized (arrow).

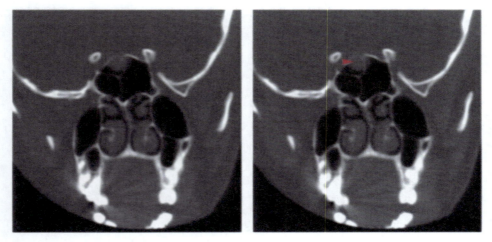

Fig. 80 CT scan (bone window). Coronal slice. The tumor has invaded the ONODI cell (arrow).

II. Preoperative management
II.A. Neuroradiological investigations
II.A.1 CT

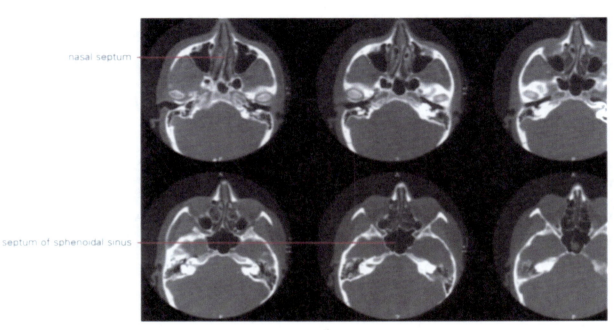

Fig. 81 CT scan. Thin serial axial slices showing the nasal conchae, the antero-posterior and lateral orientation of the nasal septum and the septum of the sphenoidal sinus, the antero-posterior and lateral relationship between ethmoidal and sphenoidal sinus.

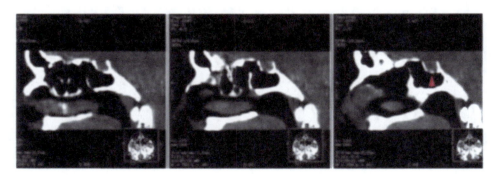

Fig. 82 CT scan. Serial sagittal slices showing the antero-posterior relationship between the ethmoidal and the sphenoidal sinus and a partial erosion of the floor of the sella turcica (arrow).

II. Preoperative management

II.A. Neuroradiological investigations

II.A.2 MRI

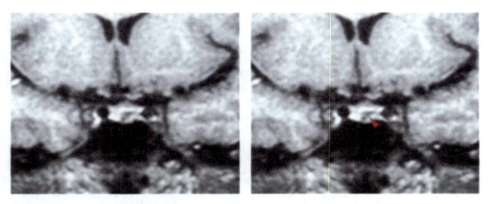

Fig. 83 MRI, T1, coronal slice after i.v. administration of GADOLINIUM. A pituitary microadenoma (arrow) is identified on the left side.

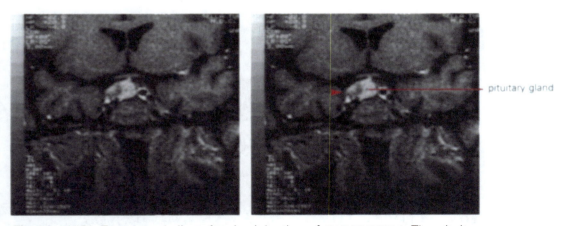

Fig. 84 MRI, T1, coronal slice after i.v. injection of GADOLINIUM. The pituitary adenoma is on the right side (arrow).

II. Preoperative management
II.A. Neuroradiological investigations

II.A.2 MRI

fornix displaced
(yellow arrows)

intracavernous ICA
(red arrows)

extension into the
sphenoidal sinus
(white arrow)

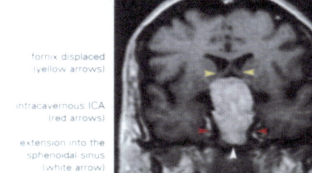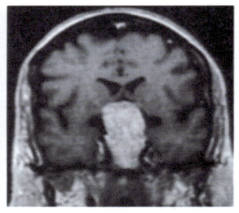

Fig. 85 MRI, T1, coronal slice after i.v. GADOLINIUM injection. Intra- and suprasellar pituitary adenoma. The coronal slices are particularly useful to visualise the lateral extension of the lesion. Note the internal carotid arteries (red arrows). In spite of the large size of the tumor, the internal carotid arteries are not displaced laterally. The tumor goes up to third ventricle, (the fornices are dislocated, yellow arrow) and down to the sphenoidal sinus (white arrow).

fornix

sphenoidal sinus

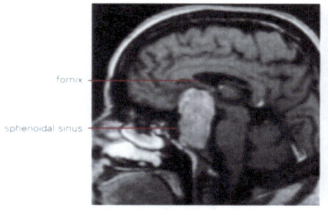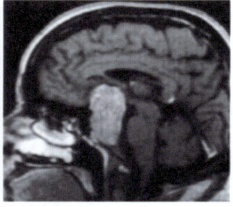

Fig. 86 MRI, T1, sagittal slice after i.v. GADOLINIUM injection. The sagittal slices are particularly useful to visualise the anterior and posterior extension of the tumor. The lesion occupies the postero-inferior part of the sphenoidal sinus; there is no extension anteriorly or posteriorly to the sella turcica. In the suprasellar area the tumor extends up to third ventricle.

II. Preoperative management

II.A. Neuroradiological investigations

II.A.2 MRI

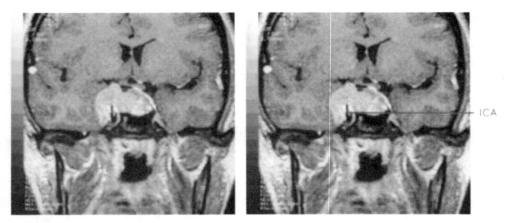

Fig. 87 MRI, T1, coronal slice after i.v. GADOLINIUM injection. Intra-suprasellar GH-PRL macroadenoma with extension into the right cavernous sinus. This pituitary adenoma involves both medial and lateral compartments of the cavernous sinus and incorporates the internal carotid artery (ICA).

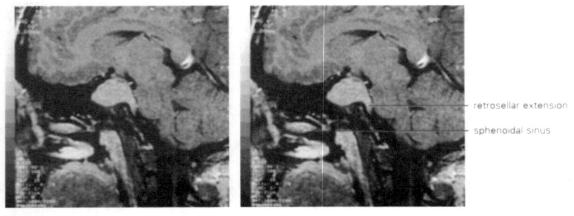

Fig. 88 MRI, T1, sagittal slice after i.v. GADOLINIUM injection. Note the posterior extension into the interpeduncular cistern. The sphenoidal sinus is marginally involved.

II. Preoperative management
II.A. Neuroradiological investigations

II.A.2 MRI

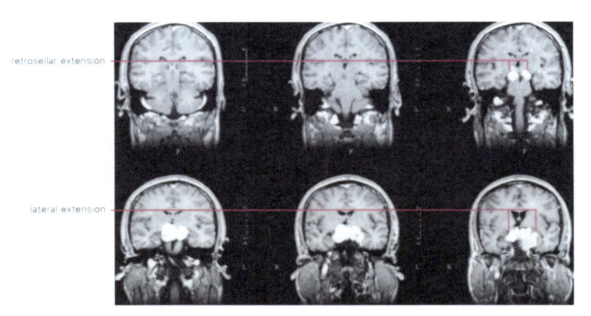

Fig. 89 MRI T1, coronal serial slices after i.v. GADOLINIUM injection. Giant macroadenoma with involvement of both cavernous sinuses, the suprasellar compartment and the sphenoidal sinus.

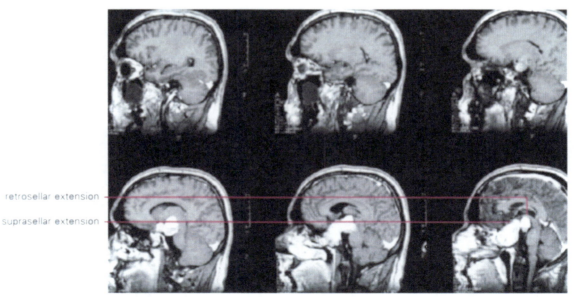

Fig. 90 MRI T1, sagittal serial slices after i.v. GADOLINIUM injection. The sagittal slices better define the antero-posterior, supra- and infrasellar extension.

II. Preoperative management

II.B. Operating theatre

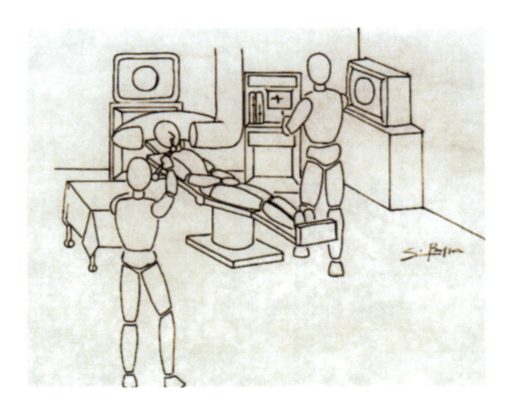

II. Preoperative management

II.B. Operating theatre

II.B.1. Positioning of the patient

In transsphenoidal surgery the position of the patient on the operating table depends on the experience and preference of the surgeon.

For this approach we have chosen a position that we called the **"cradle-like position"**:

- Patient supine
- Gentle Trendelenburg
- Trunk a little raised in relation to the hips
- Lower limbs gently lifted with the knees flexed
- Head slightly hyperextended and turned towards the surgeon, who is on the patient's right

The reasons for our choice are based on:
- Positive assessment by the anesthesiologist
- No risk of air embolism
- No need for the Mayfield headrest
- Comfortable working position for the surgeon
- Spontaneous downflow of blood and irrigation fluids
- Favorable projection to dissect the suprasellar part of the lesion
- Risk of endoscope or of other instruments falling into the operating field due to their inclination toward the bottom, as in the pure supine position minimized.

A C-arm lateral fluoroscopic device is positioned. This is useful mainly in the first procedures, in the exposure of the sphenoidal sinus ostium and the sella turcica phase.

II. Preoperative management

II.B. Operating theatre

II.B.1. Positioning of the patient

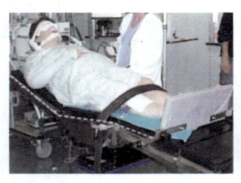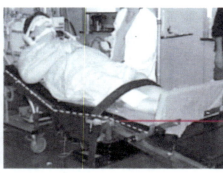

Fig. 91 Position of the patient. The patient is supine, with the trunk a little raised in relation to the hips, the lower limbs gently lifted, with the knees flexed. The head is slightly hyperextended and turned towards the surgeon. The white line indicates the operating table position we called the "cradle-like" position.

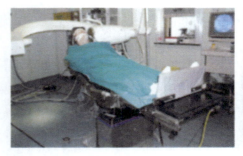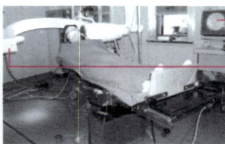

Fig. 92 The C-arm is positioned. The use of this device is decreasing and in our opinion it will be eliminated with experience. Nowadays a valid alternative is the neuronavigator.

II. Preoperative management
II.B. Operating theatre
II.B.1. Positioning of the patient

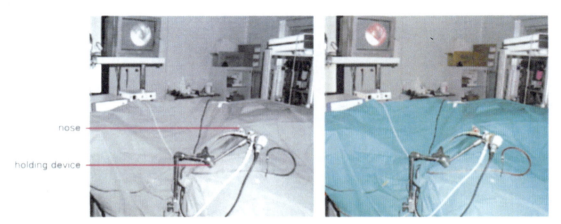

Fig. 93 Only the nose can be seen and the holding device can be employed from the beginning of the prodecure.

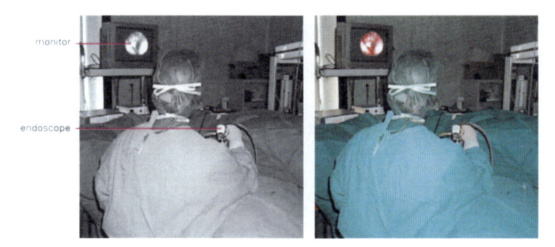

Fig. 94 During the whole procedure the surgeon, the anesthesiologist and the nurse will look at the monitor.

II. Preoperative management

II.B. Operating theatre

II.B.2. Equipment

Operating without the use of a nasal speculum, which gives no surgical field beyond its two blades, the procedure is carried out through only one nostril, using a 4 mm rigid diagnostic endoscope as the sole visualizing tool. The endoscope is inserted in a rigid sleeve 5.6 mm in diameter and connected to a cleaning-irrigation system. The instruments are introduced, close and parallel to the endoscope, into the same nostril. The endonasal operation begins with the enlargement of the sphenoidal sinus ostium without septum transfixion. The choice of the most suitable nostril and the sufficient enlargement of the sphenoidal sinus ostium removes the necessity to use both nostrils. The endoscope is fixed to an endoscope holder. This can be moved at any time, providing the surgeon with a stable image of the surgical field and leaving both hands free to hold the suction with the left hand and another instrument, such as a curette, with the right.

II. Preoperative management

II.B. Operating theatre

II.B.2. Equipment

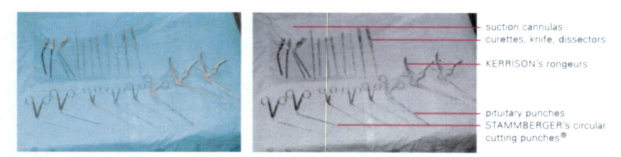

Fig. 95 Even though the instrumentation is very varied, the instruments most used during the procedure are shown.

- suction cannulas
- curettes, knife, dissectors
- KERRISON's rongeurs
- pituitary punches
- STAMMBERGER's circular cutting punches®

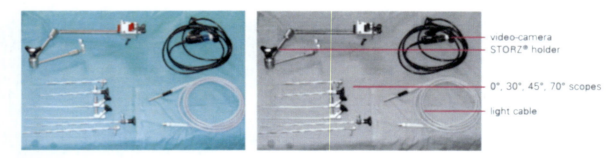

Fig. 96 The endoscopes, the light cable, the video-camera, the holding device.

- video-camera
- STORZ® holder
- 0°, 30°, 45°, 70° scopes
- light cable

II. Preoperative management

II.B. Operating theatre

II.B.2. Equipment

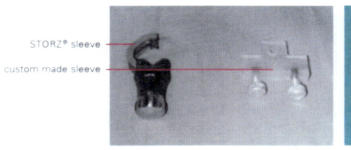

Fig. 97 The sleeve attached to the tip of the holding device. Left: STORZ® sleeve, right: our custom made sleeve.

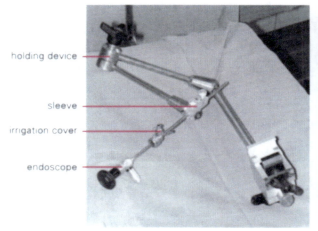
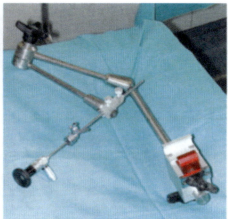

Fig. 98 The endoscope is inserted into a cover for irrigation. The cover is introduced through our custom made sleeve, attached to the tip of the STORZ® holding device. This system allows the surgeon to extract and clean the endoscope, while the cover attached to the holder remains inside the operating field.

II. Preoperative management
II.B. Operating theatre

II.B.2. Equipment

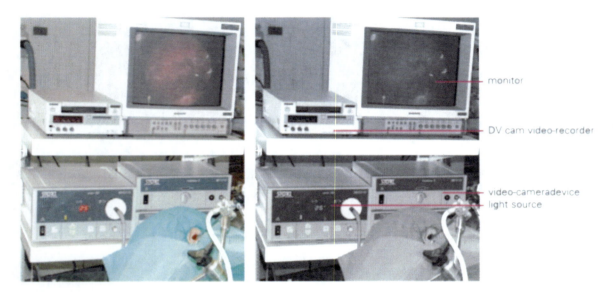

Fig. 99 STORZ® equipment.

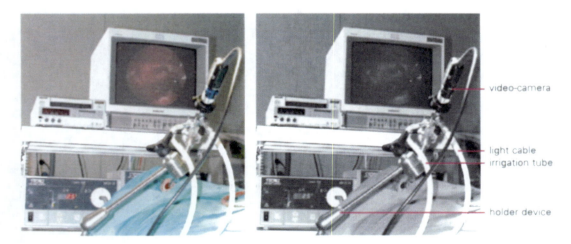

Fig. 100 The video-camera, the light source and an irrigation system have been installed.

II. Preoperative management
II.B. Operating theatre
II.B.2. Equipment

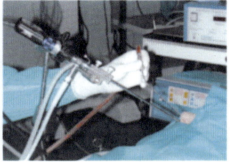

Fig. 101 The endoscope is fixed to the holder. The suction cannula is introduced along and coaxial to the endoscope.

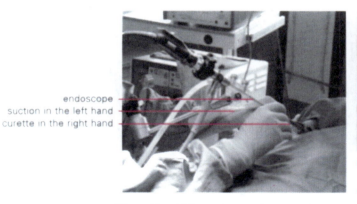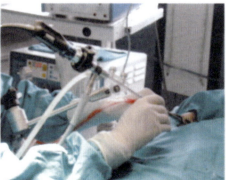

Fig. 102 When the endoscope is secured to the holder, both hands are easily employed.

III. Surgical procedure

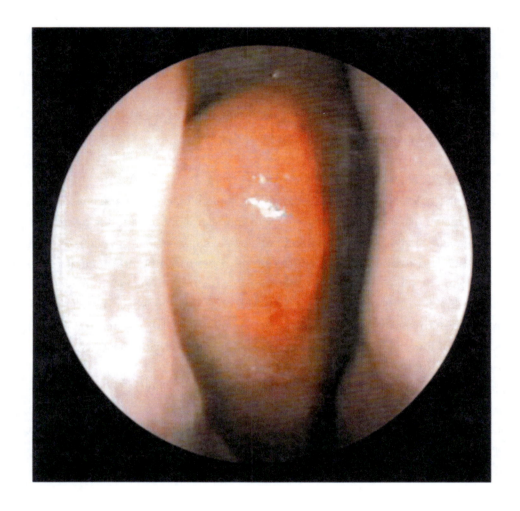

III. Surgical procedure

III.A. Surgical steps

Disinfection of the nasal cavities and decongestant treatment using contact anesthetics are carried out. We eliminated the infiltration of anesthetics because this manoeuvre often gives rise to a little bleeding.

The procedure can be divided into 4 phases:

• **1ˢᵗ step** (nasal phase):
Due to the oblique latero-medial trajectory of the approach (the parasellar growth of the lesion usually suggests the entry of the nostril opposite to the site of the lateral expansion) and after the neuroradiological studies (anatomy of inner nose and sphenoid region) left or right access is determined before beginning the procedure. In any event with the endoscope it is correct to inspect both sides to determine the most suitable approach anatomically. The space between the nasal septum and the middle nasal concha may initially be very narrow. Lateral luxation of the middle nasal concha may be needed to create space to push the endoscope forward. If a concha bullosa is present, the head of the middle nasal concha can be removed. When sufficient space is obtained, the endoscope is advanced along the inferior margin of the middle nasal concha to reach the choana and the spheno-ethmoidal recess. The sphenoidal sinus ostium is found in the upper half of the spheno-ethmoidal recess.

• **2ⁿᵈ step** (natural ostium enlargement):
The holder is mounted to keep the image fixed and to allow the surgeon to work with both hands. The mucosa around the ostium is coagulated and the ostium widened to a diameter of 15-20 mm. The sphenoidal rostrum is removed. This is crucial as it permits a very wide working angle to accommodate the endoscope plus two instruments. It also provides the exact trajectory needed to remove the tumor, particularly in the case of supra and/or parasella turcica expansion of the lesion. The ostium should be enlarged opposite to the direction taken by the surgeon at the level of the sella turcica (corresponding to the key hole concept).

III. Surgical procedure

III.A. Surgical steps

• **3rd step** (preparation of the sphenoidal sinus and opening of the sella turcica):
The sphenoid mucosa is moved aside as much as is necessary for opening the sella turcica. It is not removed if it is not infiltrated in order to preserve its ciliary function. The septum or septa of the sphenoidal sinus are removed. In a well pneumatized sinus, a very wide panoramic view of the posterior and lateral walls of the sphenoidal sinus is obtained and the sphenoid plane, the sella turcica and the clivus are found at the posterior wall of the sphenoidal sinus in a cranio-caudal direction. The optic protuberance, the optico-carotid recess and the carotid protuberance are found at the lateral walls of the sphenoidal sinus in a rostro-caudal direction. The floor of the sella turcica is opened using a high speed drill if it has a hard consistency; otherwise micro KERRISON or STAMMBERGER cutting punches can be used. The dura is incised.

• **4th step** (tumor removal):
Curettage and aspiration of the lesion is performed using slow, circular movements in all directions. Fluoroscopy is used to control the extent of the removal and to indicate the presence of air in or over the sella turcica if there is a CSF leak. The angled endoscopes (30, 45 and 70 degrees) are very useful for the verification of the supra- and parasellar extension of a lesion. At the end of the procedure the endoscope is introduced further on and an inspection is made of the cistern and the pituitary gland, if visible. Finally, sella turcica reconstruction is performed according to the the standard guidelines. *No nasal packing is used at the end of the procedure. Always work under direct visual control!*

The limits of the procedure can be summarised as follows:
• A basic knowledge of the trans-sphenoidal approach is required and at least 10–12 procedures should be performed to become fully acquainted with the technique. Skill is needed to insert the endoscope plus two instruments into the small opening of the nostril and then glide along it to the target in the sella turcica. However, on the other hand to manoeuvre has a very wide angle of action, in the absence of the nasal speculum.

III. Surgical procedure

III.A. Surgical steps

- The endoscope does not provide three-dimensional stereoscopic images and there is a "barrel-type" distortion of the image which can lead to misinterpretation of the distance of an object, particularly its borders.

- Absence of specific instrumentation. If appropriate, decongestant treatment of the nasal mucosa is performed and the sphenoidal ostium is adequately enlarged. The operation can be conducted with the instruments currently employed in the traditional trans-sphenoidal microsurgical approach and for endoscopic sinus surgery.

- This problem will be overcome with the production of a specific series of tools that we are in the process of developing with a view to improve the execution of the technique and devising new surgical perspectives. The architect LE CORBUSIER said in his book "Towards a new architecture": *In every field of industry new problems have presented themselves and new tools have been created capable of resolving them. If this new fact be set against the past, then you have revolution.*

The advantages of this technique are:
- It is not necessary to use the nasal speculum, thus surgical trauma is reduced due to the absence of the spreading of the blades of the transsphenoidal retractor.
- Better visualisation of the structures of the sella turcica (interface pituitary gland tumor, cistern tumor).
- The use of angled endoscopes provides better visual and operative control of supra- and parasellar areas.
- More radical removal of large lesions and better inspection of remote areas.
- Access to recurrences is easier and safer because of the previous sphenoidectomy and the more complete and wider anatomical view.
- Reduction of length ond cost of hospital stay.

III. Surgical procedure
III.A. Surgical steps

III.A.1. Endonasal approach to the sphenoidal sinus ostium

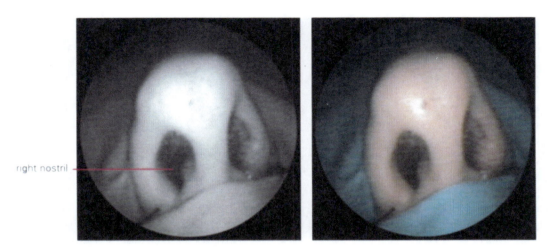

Fig. 103 0° scope. The nose before starting the procedure. The surgeon is about to introduce the endoscope into the right nostril.

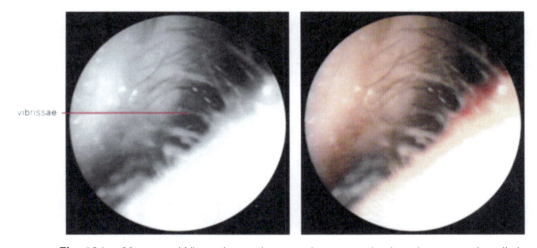

Fig. 104 0° scope. When the endoscope is approached to the nose, the vibrissae are seen very clearly.

III. Surgical procedure

III.A. Surgical steps

III.A.1. Endonasal approach to the sphenoidal sinus ostium

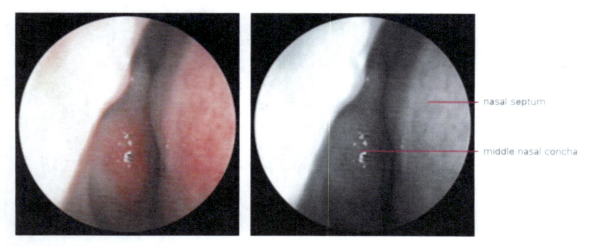

Fig. 105 0° scope. Exploration of the right nostril.

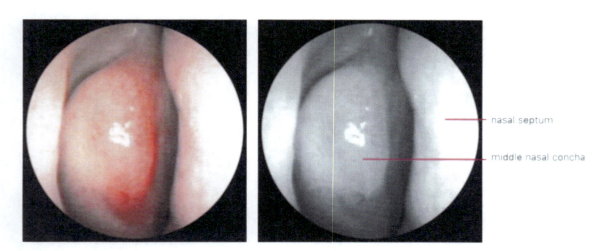

Fig. 106 0° scope. Note the large middle nasal concha (concha bullosa). The space between the middle nasal concha and the nasal septum is very narrow. This condition does not allow sufficient room to work, so it might be necessary to cut the tip of the middle nasal concha.

III. Surgical procedure
III.A. Surgical steps

III.A.1. Endonasal approach to the sphenoidal sinus ostium

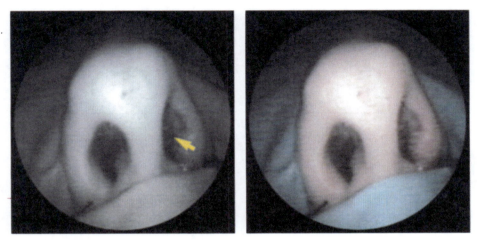

Fig. 107 0° scope. The arrow indicates the direction and the site of entrance of the endoscope.

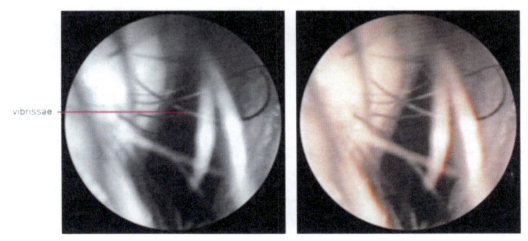

Fig. 108 0° scope. The endoscope is approximated to the left nostril and the nasal vibrissae are identified

III. Surgical procedure

III.A. Surgical steps

III.A.1. Endonasal approach to the sphenoidal sinus ostium

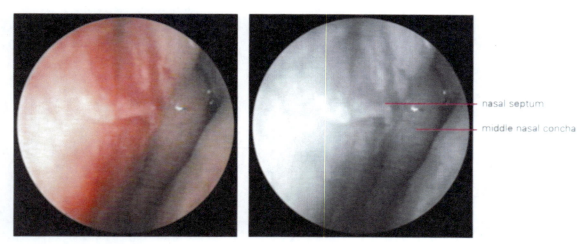

Fig. 109 0° scope. When the endoscope is further introduced into the left nostril, the middle nasal concha is identified.

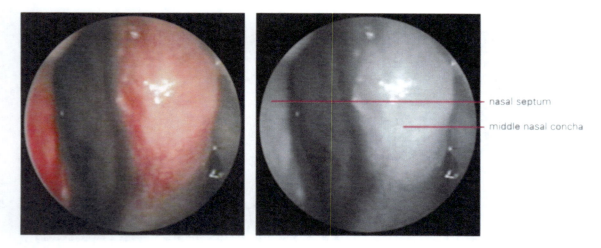

Fig. 110 0° scope. The middle nasal concha is large, but there is no concha bullosa and, due to the deviation of the nasal septum, there is more space than in the right side. We chose the left nostril to perform the procedure. We usually the middle nasal concha do not luxate, except when the space between the middle turbinate and the nasal septum is very narrow.

III. Surgical procedure

III.A. Surgical steps

III.A.1. Endonasal approach to the sphenoidal sinus ostium

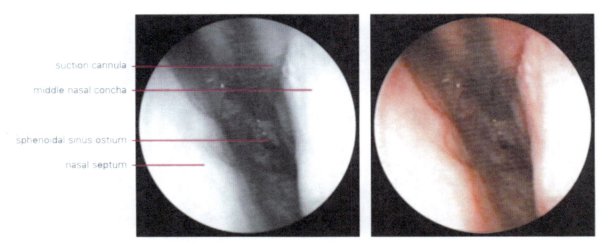

Fig. 111 0° scope. The further introduction of the endoscope makes it possible to identify the sphenoidal sinus ostium at the bottom. Beyond the endoscope, the suction cannula and another instrument can be used.

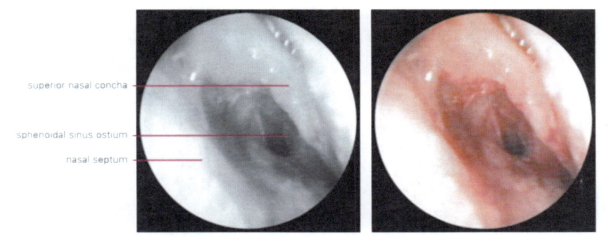

Fig. 112 0° scope. The endoscope is introduced close to the sphenoidal sinus ostium and, as the endoscope can now be fixed to the holder device, it is possible to use both hands.

III. Surgical procedure

III.A. Surgical steps

III.A.1. Endonasal approach to the sphenoidal sinus ostium

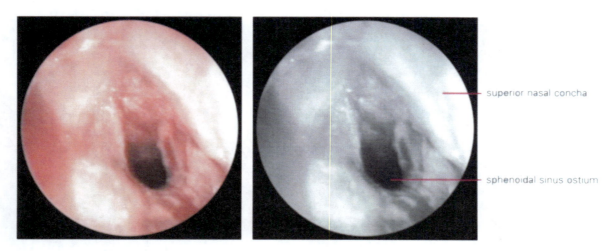

Fig. 113 0° scope. Closer view of the sphenoidal sinus ostium.

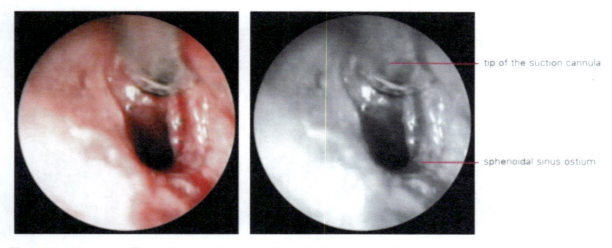

Fig. 114 0° scope. The sphenoidal sinus ostium is the crucial point, where this surgical procedure is started.

III. Surgical procedure
III.A. Surgical steps

III.A.2. Enlargement of the sphenoidal sinus ostium

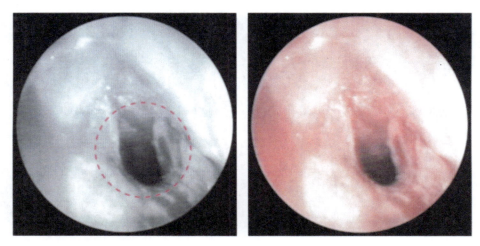

Fig. 115 0° scope. When the ostium is identified, cauterization of the mucosa around the ostium is performed (dotted line).

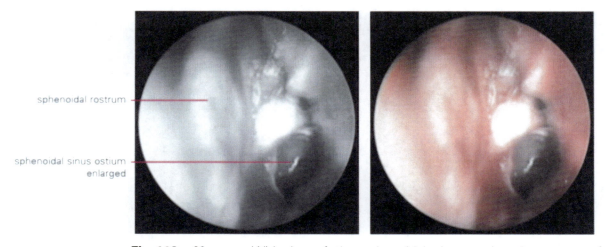

Fig. 116 0° scope. Widening of the sphenoidal sinus ostium by means of KERRYSON's rongeurs.

III. Surgical procedure

III.A. Surgical steps

III.A.2. Enlargement of the sphenoidal sinus ostium

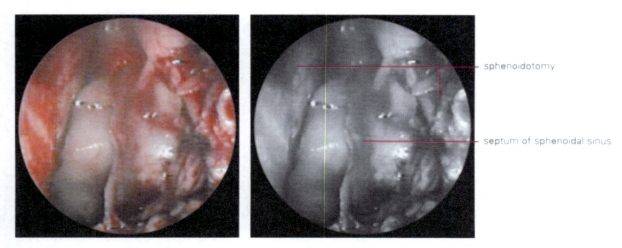

Fig. 117 0° scope. Sphenoidotomy. The endoscope is positioned outside the sphenoidal sinus.

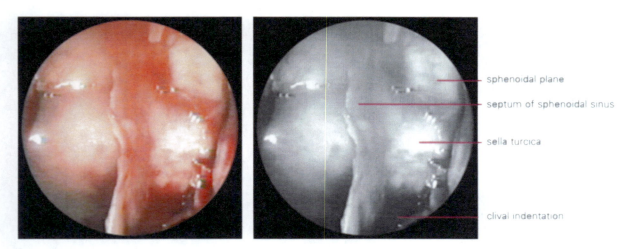

Fig. 118 0° scope. Endoscope inside the sphenoidal sinus.

III. Surgical procedure
III.A. Surgical steps
III.A.3. Preparation of the sphenoidal sinus

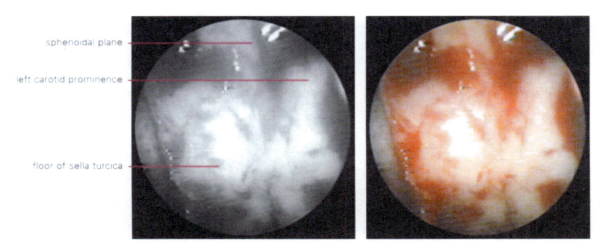

Fig. 119 0° scope. The septum of the sphenoidal sinus has been removed.

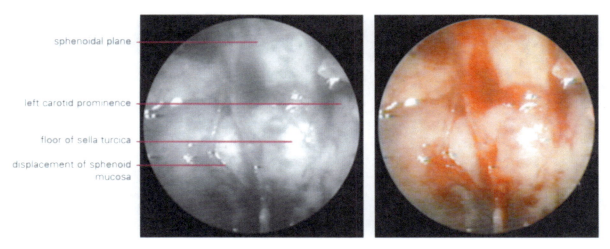

Fig. 120 0° scope. Gentle displacement of the mucosa covering the floor of the sella turcica.

III. Surgical procedure

III.A. Surgical steps

III.A.3. Preparation of the sphenoidal sinus

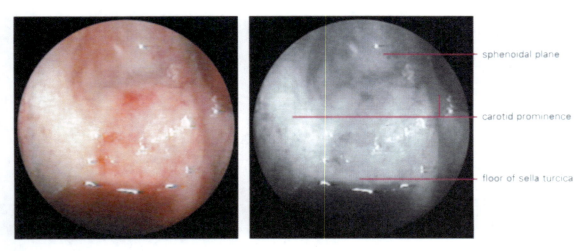

Fig. 121 0° endoscope inside the sphenoidal sinus. Panoramic view.

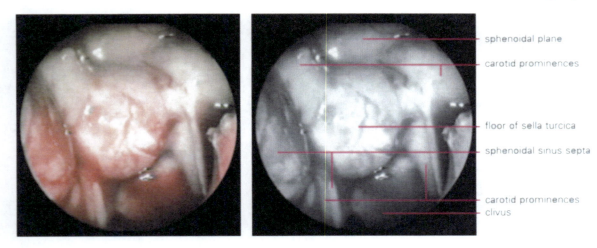

Fig. 122 0° endoscope inside the sphenoidal sinus. The floor of the sella turcica is enlarged. A paramedian right septum is still present in this multi-septated sphenoidal sinus.

III. Surgical procedure
III.A. Surgical steps

III.A.4. Opening of the floor of the sella turcica

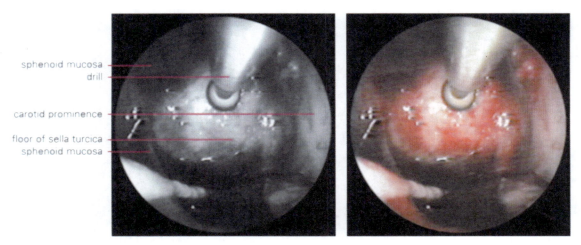

Fig. 123 0° scope. Thick floor of the sella turcica opened by means of a microdrill.

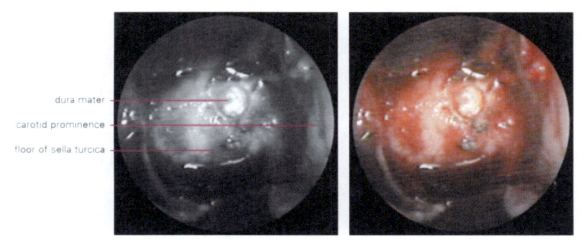

Fig. 124 0° scope. The first hole in the floor of the sella turcica has been made.

III. Surgical procedure

III.A. Surgical steps

III.A.4. Opening of the floor of the sella turcica

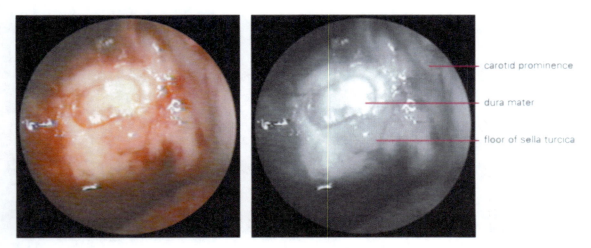

Fig. 125 0° scope. A second hole has been made through which instruments are introduced and the removal of the floor of the sella turcica is completed.

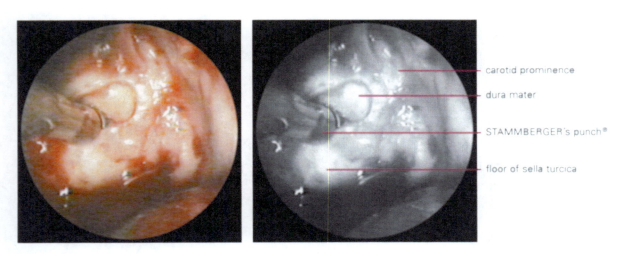

Fig. 126 0° scope. Enlargement of the holes in the floor of the sella turcica by means of the STAMMBERGER circular cutting punch®.

III. Surgical procedure
III.A. Surgical steps
III.A.4. Opening of the floor of the sella turcica

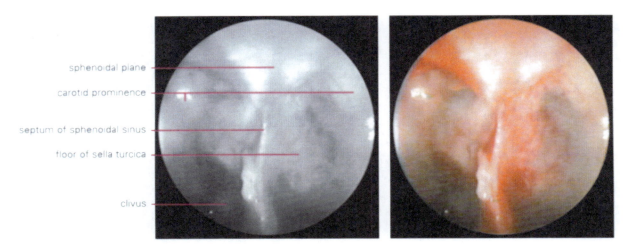

Fig. 127 0° scope. In this case the floor of the sella turcica is thin and a thick median septum in the sphenoidal sinus is found.

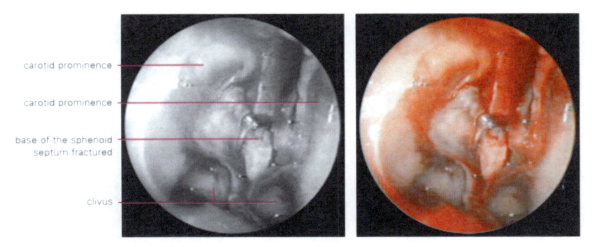

Fig. 128 0° scope. When the median septum of the sphenoidal sinus is removed, the thin floor of the sella turcica is opened.

III. Surgical procedure

III.A. Surgical steps

III.A.4. Opening of the floor of the sella turcica

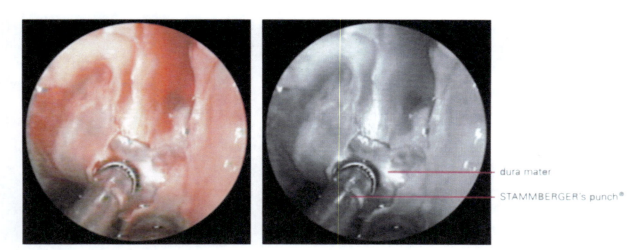

Fig. 129 0° scope. After the septum of sphenoidal sinus is removed, a circular STAMMBERGER's cutting punch® is employed in order to open the floor of the sella turcica.

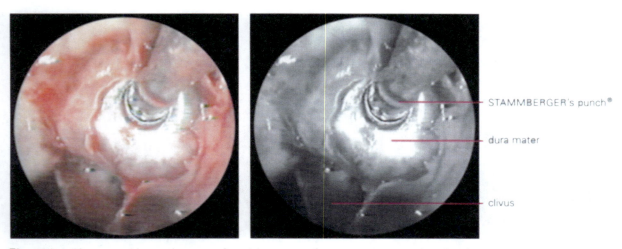

Fig. 130 0° scope. Note the comfortable use of STAMMBERGER's circular cutting punch® in all directions. The removal of the floor of the sella turcica is almost completed.

III. Surgical procedure
III.A. Surgical steps
III.A.5. Opening of the dura mater

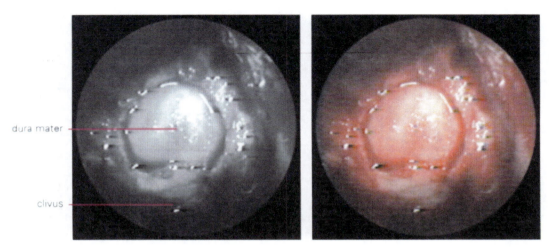

Fig. 131 0° scope. The removal of the floor of the sella turcica has been completed and the sellar dura mater exposed.

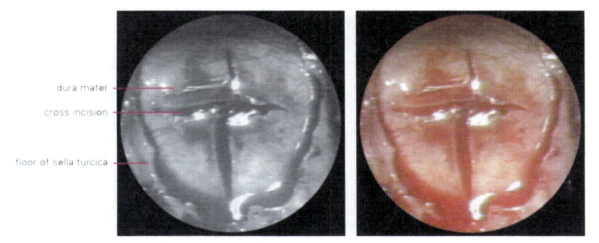

Fig. 132 0° scope. The dura mater has been opened using a cross-shaped incision. The endoscope is introduced into the sphenoidal sinus very close to the sella turcica.

III. Surgical procedure

III.A. Surgical steps

III.A.5. Opening of the dura mater

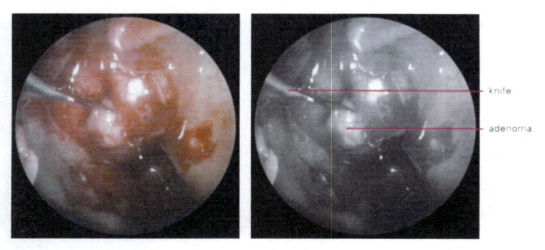

Fig. 133 0° endoscope inside the sphenoidal sinus. When the dura mater is incised with the knife, the lesion often flows out spontaneously.

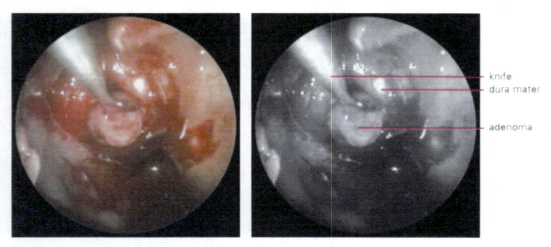

Fig. 134 0° endoscope inside the sphenoidal sinus. After the first cut the knife is directed laterally to complete the dural incision.

III. Surgical procedure
III.A. Surgical steps
III.A.6. Removal of the lesion

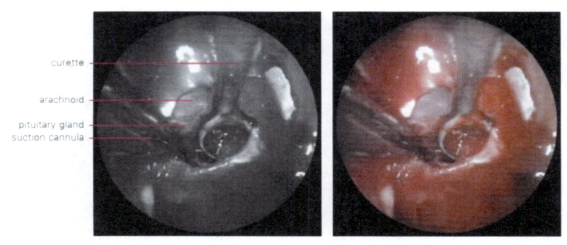

Fig. 135 45° scope inside the sphenoidal sinus. The suprasellar cistern and the pituitary gland are pushed upwards and the retrosellar part of the tumor is removed.

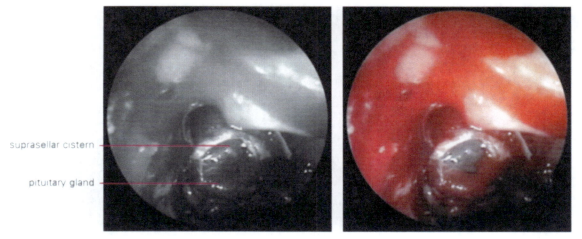

Fig. 136 45° endoscope inside the sphenoidal sinus. Closer view of the pituitary gland pushed upwards against the suprasellar cistern.

III. Surgical procedure

III.A. Surgical steps

III.A.6. Removal of the lesion

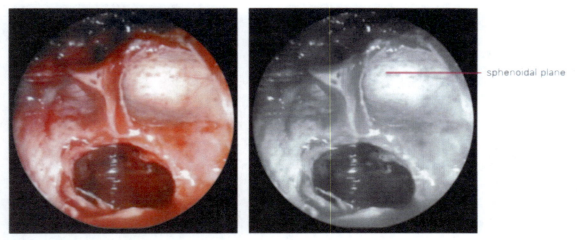

Fig. 137 45° endoscope inside the sphenoidal sinus. The sphenoidal plane is identified.

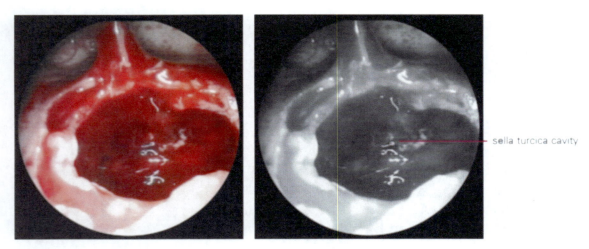

Fig. 138 45° endoscope. Closer view of the sella turcica cavity after tumor removal.

III. Surgical procedure
III.A. Surgical steps
III.A.7. Sella turcica reconstruction

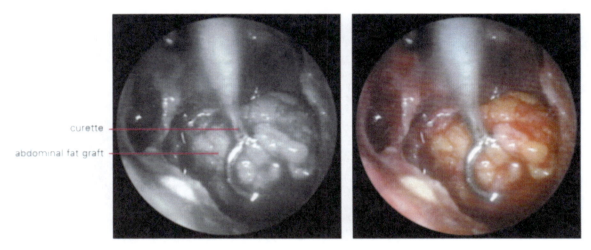

Fig. 139 0° scope. An autogenous fat graft is inserted into the sella turcica cavity.

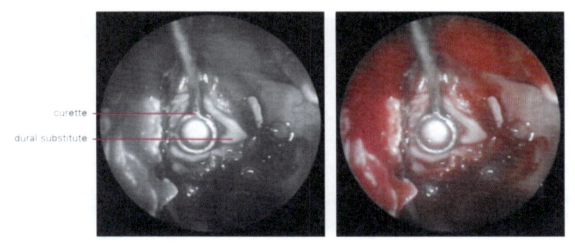

Fig. 140 0° endoscope inside the sphenoidal sinus. A dural substitute is introduced extradurally.

III. Surgical procedure

III.A. Surgical steps

III.A.7. Sella turcica reconstruction

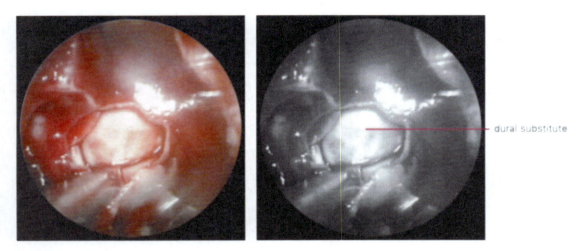

Fig. 141 0° endoscope. The dural substitute has been placed extradurally.

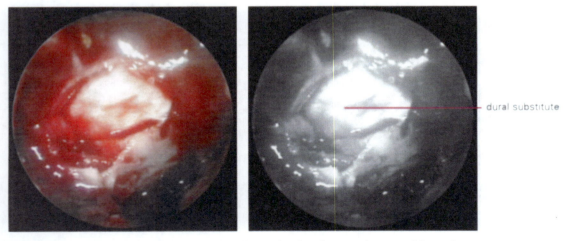

Fig. 142 0° scope. The endoscope has been further inserted to provide a closer view of the previous field. Fibrin glue is then applied and the floor of the sella turcica is reconstructed with septal sphenoidal sinus bone or titanium meshes.

Appendix

Selected clinical cases

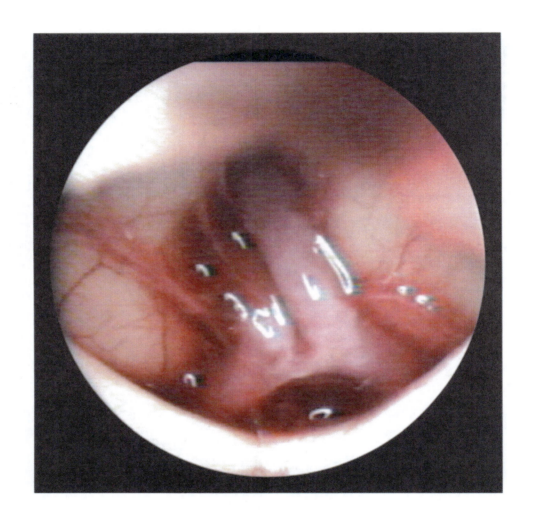

103

Appendix

Selected clinical cases

Case 1: Intra-suprasellar macroadenoma

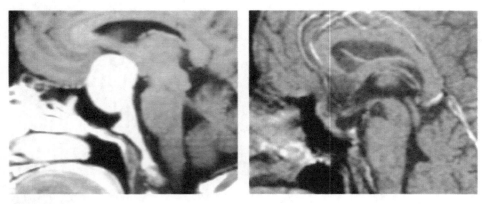

Fig. 143 MRI, sagittal slices T1 after i.v. GADOLINIUM-DTPA injection. Intra-suprasellar macroadenoma in an elderly woman before (left) and after (right) surgery.

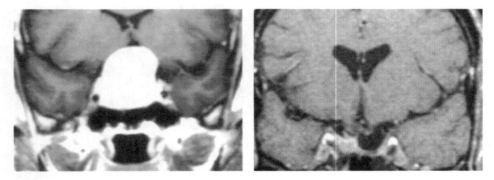

Fig. 144 MRI, T1 after i.v. GADOLINIUM-DTPA injection. Coronal view of the macroadenoma before (left) and after (right) surgery.

Appendix

Selected clinical cases

Case 1: Intra-suprasellar macroadenoma

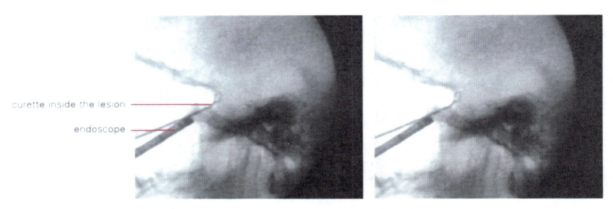

Fig. 145 Fluoroscopy during the operation. Note absence of nasal retractor and how the instrument can be moved around the endoscope.

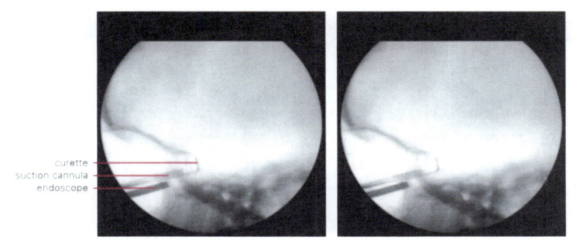

Fig. 146 Intra-operative fluoroscopy. The endoscope is fixed to the holder device and two instruments can be used together coaxial to the endoscope.

Appendix

Selected clinical cases

Case 1: Intra-suprasellar macroadenoma

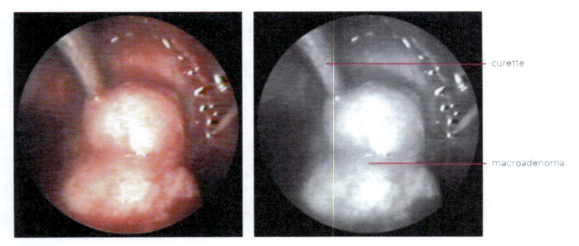

Fig. 147 0° scope. A curette is introduced in the macroadenoma.

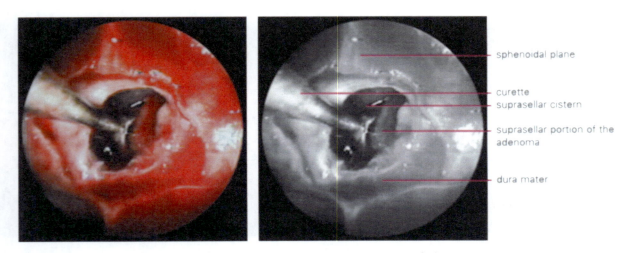

Fig. 148 0° scope. Gentle dissection of the suprasellar portion of the tumor from the suprasellar cistern. There is a high risk of CSF leakage and liquorrhea during this phase.

Appendix

Selected clinical cases

Case 1: Intra-suprasellar macroadenoma

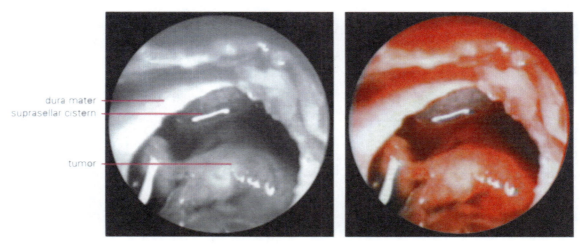

Fig. 149 0° scope. The endoscope is further introduced into the sphenoidal sinus. Closer view after downward reflection dissection of the suprasellar portion of the tumor.

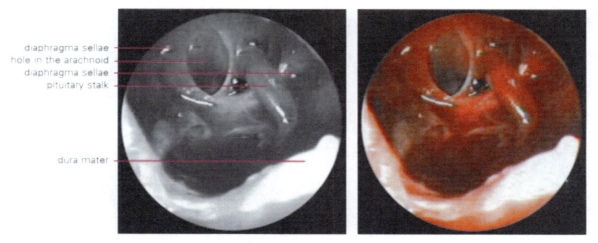

Fig. 150 0° scope. The endoscope is positioned close to the dura mater after tumor removal.

Appendix

Selected clinical cases

Case 1: Intra-suprasellar macroadenoma

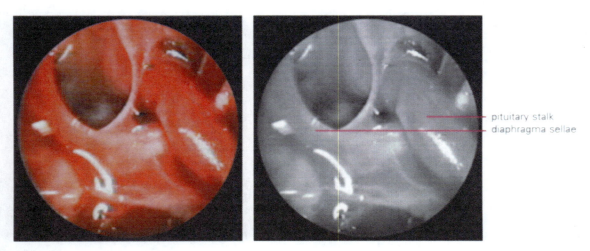

Fig. 151 The 0° endoscope has been inserted into the sella turcica and a closer view of the pituitary stalk and of the hole in the arachnoid is possible.

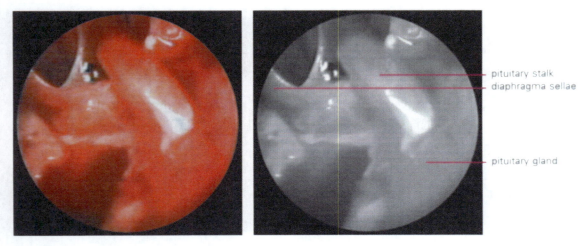

Fig. 152 30° endoscope turned to the left side inside the sella turcica. The residual pituitary gland is identified.

Appendix

Selected clinical cases

Case 2: Intra-parasellar macroadenoma

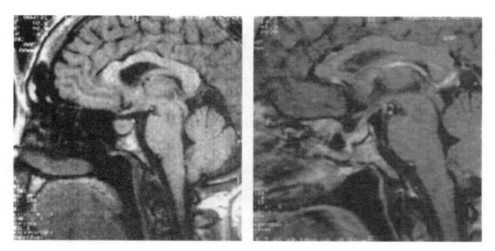

Fig. 153   MRI, sagittal slices T1 before (left) and after (right) surgery.

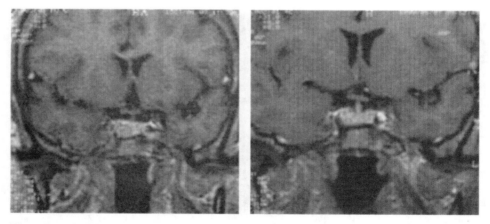

Fig. 154   MRI, coronal slices T1 after GADOLINIUM-DTPA injection. An intra-sellar pituitary adenoma on the right side before (left) and after (right) surgery.

109

Appendix

Selected clinical cases

Case 2: Intra-parasellar macroadenoma

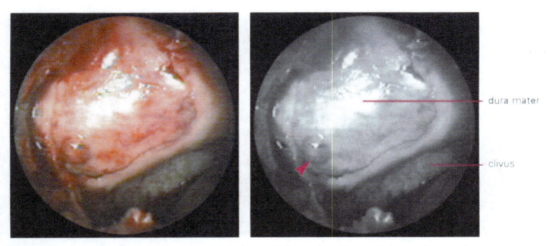

Fig. 155 0° scope in the sphenoidal sinus. When the floor of the sella turcica has been removed, the dura mater and its infiltration (arrow) by the tumor can be visualized.

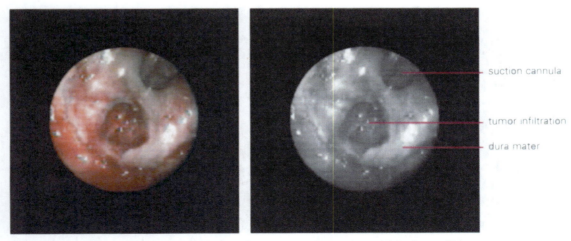

Fig. 156 Further introduction of a 0° scope into the sphenoidal sinus. Closer view of the tumor infiltrating the dura mater.

Appendix
Selected clinical cases

Case 2: Intra-parasellar macroadenoma

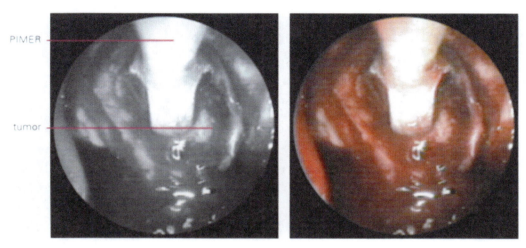

Fig. 157 0° scope. Tumor removal by a three-arm curette designed by us and called "PIMER".

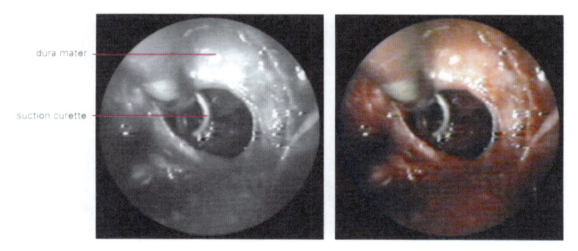

Fig. 158 0° scope. Another new curette "in action". We called it "suction curette", because it is a suction cannula with a curette on its tip.

Appendix

Selected clinical cases

Case 2: Intra-parasellar macroadenoma

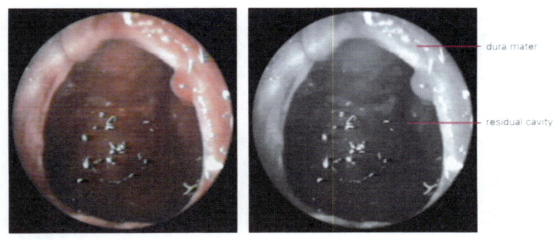

Fig. 159 0° scope. Residual cavity after tumor removal (0° endoscope closer to the dura).

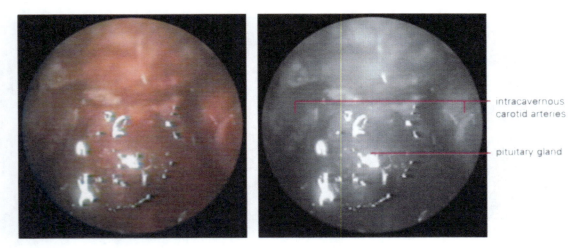

Fig. 160 0° scope inside the sella turcica after tumor removal.

Appendix

Selected clinical cases

Case 3: Solid intra-suprasellar craniopharyngeoma

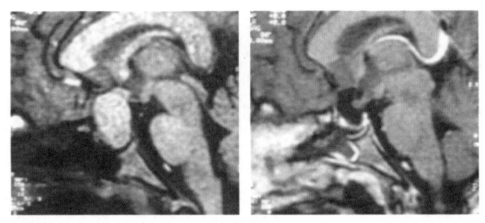

Fig. 161 T1 MRI, sagittal slices after i.v. GADOLINIUM-DTPA injection. Solid intra-suprasellar craniopharyngeoma before (left) and after (right) surgery.

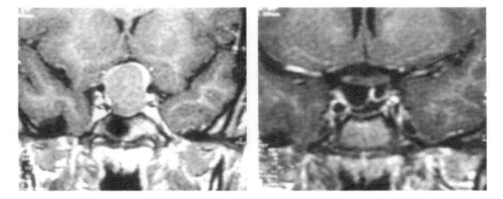

Fig. 162 T1 MRI, after i.v. GADOLINIUM-DTPA injection. The coronal slices show an solid intra-suprasellar craniopharyngeoma before surgery (left). After surgery (right) the lesion is completely removed.

Appendix

Selected clinical cases

Case 3: Solid intra-suprasellar craniopharyngeoma

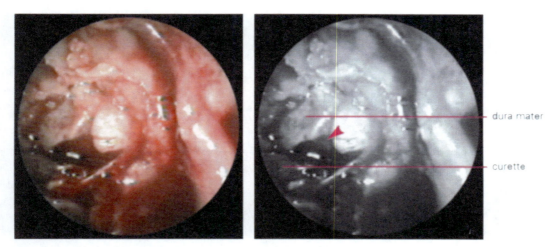

Fig. 163 0° scope. When the dura mater has been incised, the pituitary gland (arrow) can been seen. By introducing the curette inferiorly and behind the gland, as shown on MRI, tissue looking like craniopharyngeoma was removed.

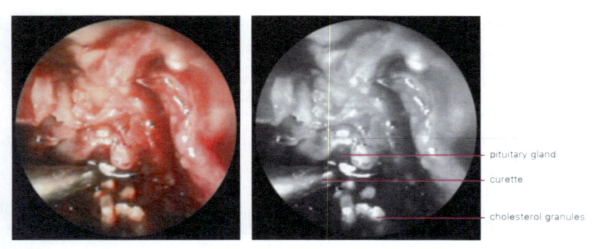

Fig. 164 0° scope. Cholesterol granules typical of a craniopharyngeoma.

Appendix
Selected clinical cases

Case 3: Solid intra-suprasellar craniopharyngeoma

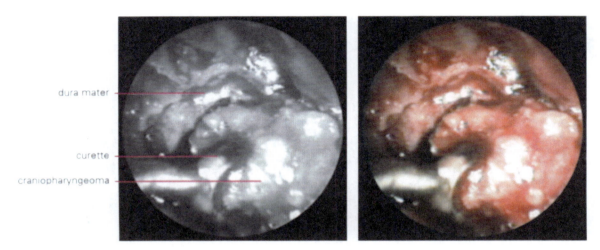

Fig. 165 0° scope. Curettage of the craniopharyngeoma.

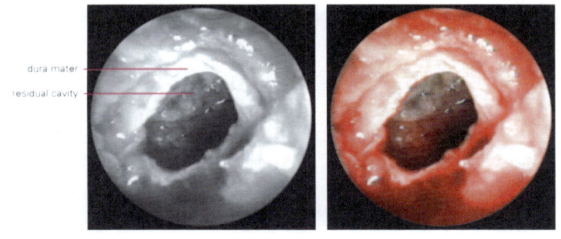

Fig. 166 0° scope in the sphenoidal sinus. Residual cavity after craniopharyngeoma removal.

Appendix

Selected clinical cases

Case 3: Solid intra-suprasellar craniopharyngeoma

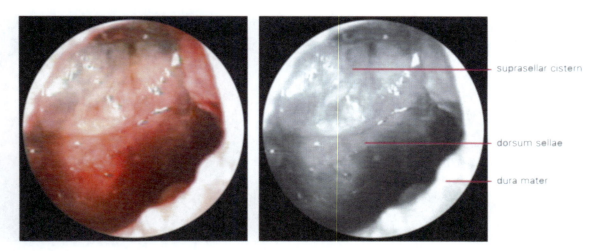

Fig. 167 0° endoscope approximated to the sella turcica cavity after tumor removal.

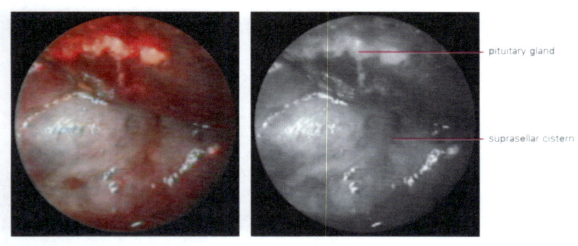

Fig. 168 30° endoscope turned upwards inside the sella turcica cavity, in order to see the pituitary gland gradually coming down.

Appendix

Selected clinical cases

Case 4: Cystic intra-suprasellar craniopharyngeoma

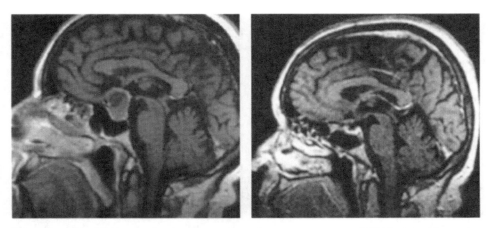

Fig. 169 T1 MRI, sagittal slices after i.v. GADOLINIUM-DTPA injection. An cystic intra-suprasellar craniopharyngeoma in a 70-year-old female before (left) and after (right) surgery.

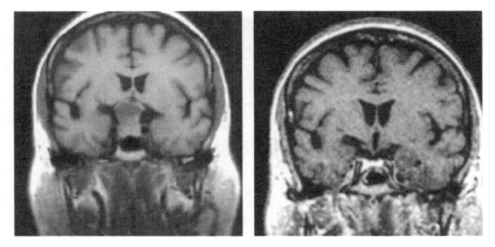

Fig. 170 T1 MRI, coronal slices after i.v. GADOLINIUM-DTPA injection. The craniopharyngeoma before (left) and after (right) surgery.

Appendix

Selected clinical cases

Case 4: Cystic intra-suprasellar craniopharyngeoma

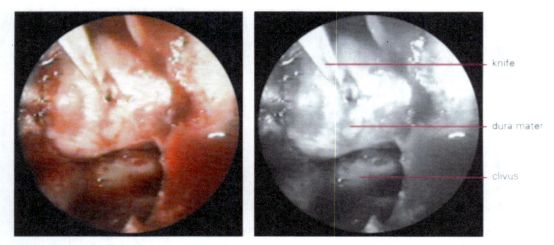

Fig. 171 0° endoscope in the sphenoidal sinus. The knife is cutting the dura mater.

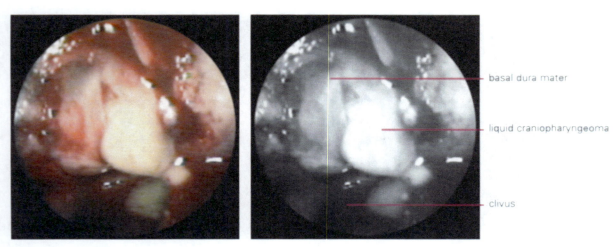

Fig. 172 0° endoscope. The "cream-like" liquid content of the craniopharyngeoma is flowing out spontaneously.

Appendix

Selected clinical cases

Case 4: Cystic intra-suprasellar craniopharyngeoma

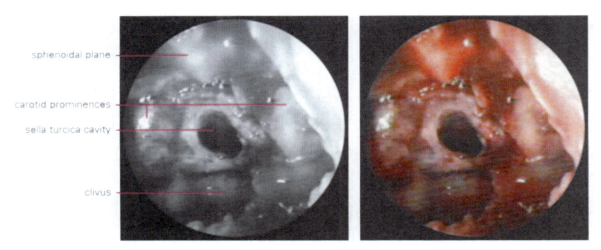

Fig. 173 0° scope outside the sphenoidal sinus. Panoramic view of the sella turcica region after tumor removal.

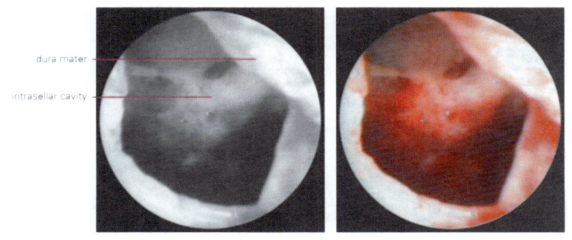

Fig. 174 0° scope inside the sphenoidal sinus close to the dura mater. The residual cavity is inspected.

Appendix

Selected clinical cases

Case 4: Cystic intra-suprasellar craniopharyngeoma

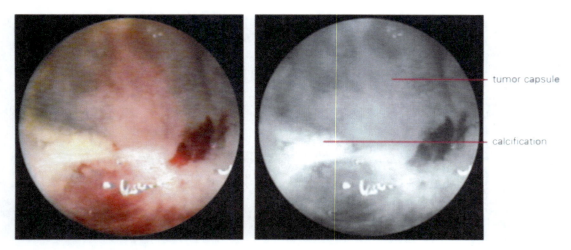

Fig. 175 0° scope further introduced to the sella turcica cavity. The tumor capsule and a calcification can be seen at close distance.

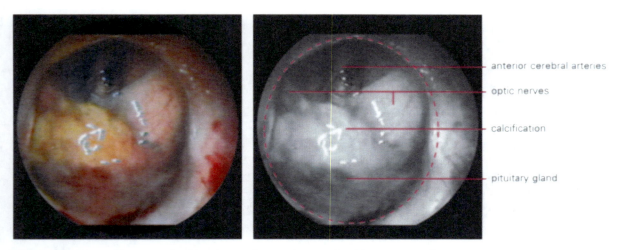

Fig. 176 70° endoscope inside the sella turcica cavity. The suprasellar structures can be seen very well because of an enlarged diaphragma sellae (dotted line).

Appendix

Selected clinical cases

Case 5: Arachnoid intra-suprasellar cyst

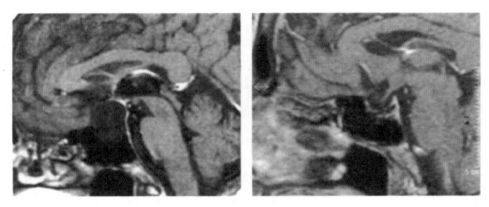

Fig. 177 T1 MRI, sagittal slices after i.v. GADOLINIUM-DTPA injection. Intra-suprasellar arachnoid cyst before (left) and after (right) surgery.

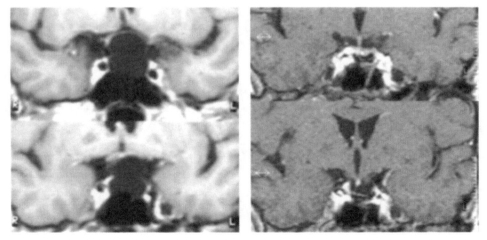

Fig. 178 T1w MRI coronal slices after GADOLINIUM-DTPA i.v. injection. The arachnoid cyst before (left) and after (right) surgery.

Appendix

Selected clinical cases

Case 5: Arachnoid intra-suprasellar cyst

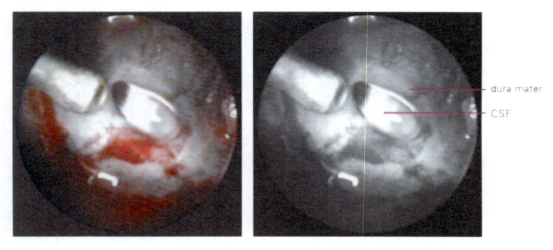

Fig. 179 0° scope in the sphenoidal sinus. The dura mater has been incised and CSF is leaking out.

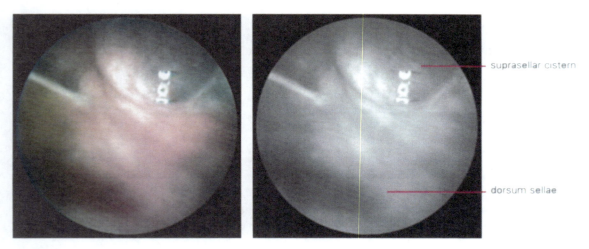

Fig. 180 0° scope inside the cyst.

Appendix

Selected clinical cases

Case 5: Arachnoid intra-suprasellar cyst

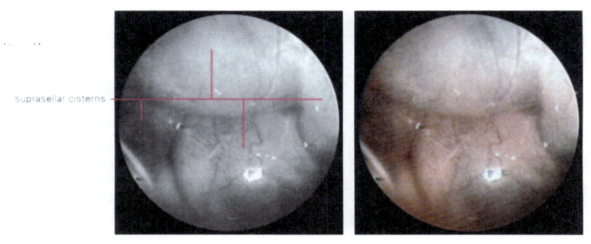

Fig. 181 30° endoscope inside the sella turcica. Note the suprasellar cisterns pushed down by a strong pulsation.

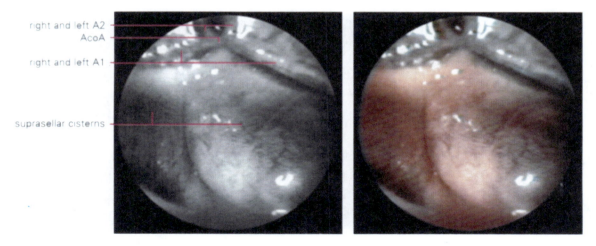

Fig. 182 30° endoscope. Endoscope kept in the same position as in the previous picture. When the suprasellar cisterns float back, the anterior circle of Willis has been identified. A1: segment A1 of anterior cerebral artery; AcoA: anterior communicating artery; A2: segment A2 of anterior cerebral artery.

Appendix

Selected clinical cases

Case 5: Arachnoid intra-suprasellar cyst

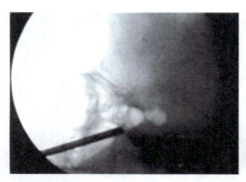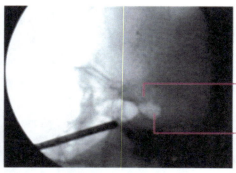

Fig. 183 Intra-operative fluoroscopy with brain pulsation sending down the suprasellar cisterns.

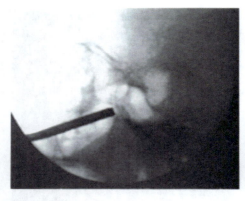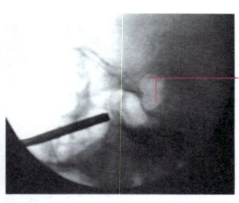

Fig. 184 Fluoroscopy with the cisterns down. Air in sella turcica and suprasellar area.

Appendix

Selected clinical cases

Case 6: Intra-suprasellar RATHKE's cleft cyst

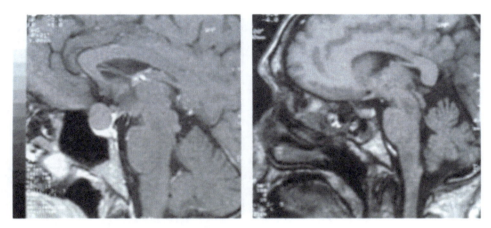

Fig. 185 T1 MRI, sagittal slices after i.v. GADOLINIUM-DTPA injection. A RATHKE's cleft cyst before (left) and after (right) surgery.

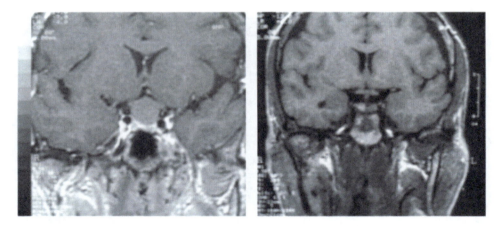

Fig. 186 T1 MRI, coronal slices after i.v. GADOLINIUM-DTPA injection. The RATHKE's cleft cyst before (left) and after (right) surgery.

Appendix

Selected clinical cases

Case 6: Intra-suprasellar RATHKE's cleft cyst

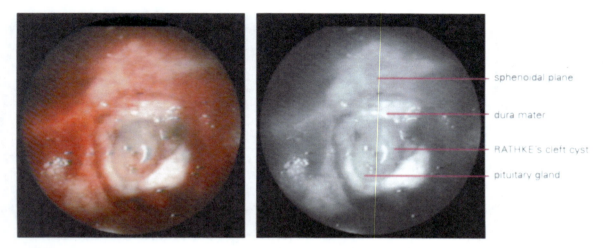

Fig. 187 0° scope inside the sphenoidal sinus. The dura mater has been removed and the RATHKE cleft cyst exposed.

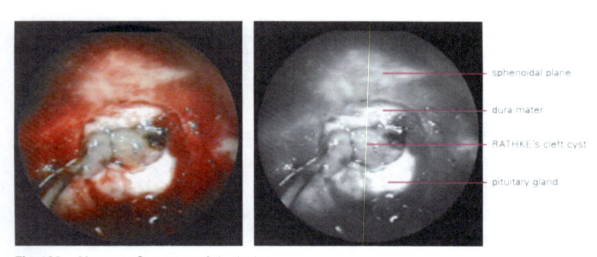

Fig. 188 0° scope. Curettage of the lesion.

Appendix

Selected clinical cases

Case 6: Intra-suprasellar RATHKE's cleft cyst

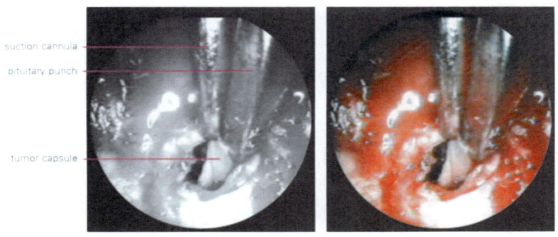

Fig. 189 0° scope inside the sphenoidal sinus. After cyst removal the capsule is dissected.

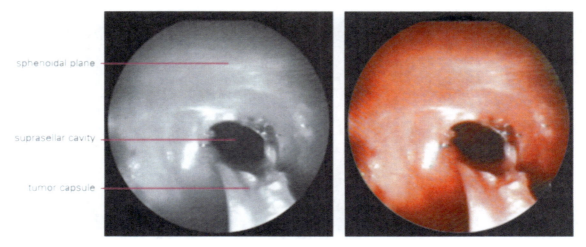

Fig. 190 0° scope. The capsule is removed.

Appendix

Selected clinical cases

Case 6: Intra-suprasellar RATHKE's cleft cyst

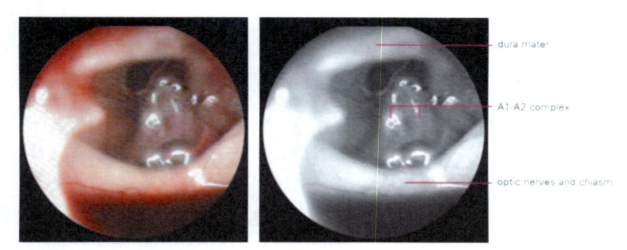

Fig. 191 45° scope. When the tumor has been removed, a 45° endoscope is introduced into the residual cavity. A1-A2: cerebral artery.

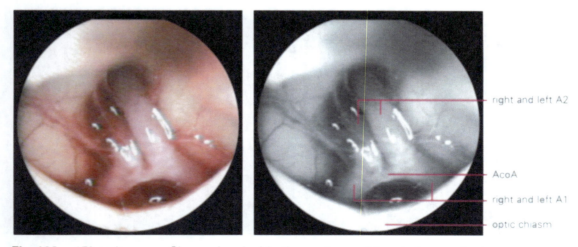

Fig. 192 45° endoscope. Closer view inside the anatomy. A1: anterior cerebral artery; AcoA: anterior communicating artery; A2: anterior cerebral artery.

References

Artico M, Pastore FS, Fraioli B, Giuffre R:
The contribution of Davide Giordano (1864–1954) to pituitary surgery: the transglabellar-nasal
approach. Neurosurgery 42 (1998): 909–912

Cappabianca P, Alfieri A, Thermes S, Buonamassa S, de Divitiis E:
Instruments for endoscopic endonasal transsphenoidal surgery.
Neurosurgery 45 (1999): 392–396

Cappabianca P, Alfieri A, de Divitiis E:
Endoscopic endonasal transsphenoidal approach to the sella turcica:
towards Functional Endoscopic Pituitary Surgery (FEPS).
Minim Invasive Neurosurg 41 (1998): 66–73

Cappabianca P, Alfieri A, Colao A, Cavallo LM, Fusco M, Peca C, Lombardi G, de Divitiis E:
Endoscopic endonasal transsphenoidal surgery in recurrent and residual pituitary adenomas:
technical note.
Minim Invasive Neurosurg 43 (2000): 38–43

Ciric I, Ragin A, Baugmgartner C, Pierce D:
Complications of the transsphenoidal surgery: results of a national survey, review of the literature,
and personal experience.
Neurosurgery 40 (1997): 225–237

Fries G, Perneczky A:
Endoscope-assisted brain surgery: Part 2–Analysis of 380 procedures.
Neurosurgery 42 (1998): 226–232

Fujii K, Chambers S M, Rhoton A L, Jr:
Neurovascular relationships of the sphenoidal sinus: a microsurgical study.
J Neurosurg 50 (1979): 31–39

Guiot G, Rougerie J, Fourestier M, FournierA, Comoy C, Vulmiere J, Groux R:
Explorations endoscopiques intracraniennes.
Press Med 71 (1963): 1225–1228

Jankowski R, Auque J, Simon C:
Endoscopic pituitary surgery.
Laryngoscope 102 (1992): 198–202

Jankovski R: Endoscopic pituitary surgery. In:
Stankiewicz JA, (ed). Advanced Endoscopic Sinus Surgery, Mosby, St. Louis (1995), pp 95–102

Jho HD, Carrau RL: Endoscopic endonasal transsphenoidal surgery: experience with 50 patients.
Neurosurgical Focus 1 (1996): 1–10

References

Jho HD, Carrau RL, Ko Y: Endoscopic pituitary surgery. In: Wilkins RH, Rengachary SS, (eds).
Neurosurgical Operative Atlas.
American Association of Neurological Surgeons, Park Ridge III (1996), pp 1–12

Jho HD: Endoscopic surgery of pituitary adenomas.
Krisht AF, Tindall GT, (eds). Pituitary disorders. Lippincott William & Wilkins, Baltimore (1999),
pp 389–403

Kennedy DW, Zenrich J, Rosenbaum AM, Johns ME: Functional endoscopic sinus surgery:
theory and diagnostic evaluation. Arch Otolaryngol Head Neck Surgery 111 (1985): 576–582

Kussmaul A: Jugenderinnerungen eines alten Arztes.
Adolf Bonz & Co. 3. Aufl., 1899, S.197

Landolt AM: History of pituitary surgery: Transsphenoidal approach.
Landolt AM, Vance ML, Reilly P, (eds). Pituitary adenomas. Churchill Livingstone,
New York (1996), pp 307–314

Lang J: Clinical Anatomy of the Nose, Nasal Cavity and Paranasal Sinuses.
Thieme, Stuttgart New York (1989)

Lang J: Hypophysial region-anatomy of the operative approaches.
Neurosurg Rev 8 (1985): 93–124

Laws ER Jr, Kern EB: Complications of transsphenoidal surgery.
Clin Neurosurg 23 (1976): 401–416

Laws ER Jr. Transsphenoidal approach to pituitary tumors.
Schmidek HH, Sweet WH, (eds). Operative Neurosurgical Techniques. WB Saunders, Philadelphia
(1995), pp 283–292

Messerklinger W: On the drainage of the frontal sinus in man.
Acta Otorhinolaryngol 63 (1967): 176–181

Rodziewicz GS, Kelley RT, Kellman RM, Smith MV:
Transnasal endoscopic surgery of the pituitary gland: technical note.
Neurosurgery 39 (1996): 189–193

Rhoton AL Jr: Microsurgical anatomy of the pituitary gland and sella turcica region. In:
Krisht AF, Tindall GT, (eds). Pituitary disorders. Lippincott William & Wilkins, Baltimore (1999),
pp 31–49

Sethi DS, Pillay PK: Endoscopic management of lesions of the sella turcica.
J Laryngol Otol 109 (1995): 956–962

References

Shikani AH, Kelly JH: Endoscopic debulking of a pituitary tumor.
Am J Otolaryngol 14 (1993): 254–256

Stammberger H: Endoscopic endonasal surgery: concepts in treatment of recurring rhinosinusitis,
part I: anatomic and pathophysiologic considerations; part II: surgical technique.
Otolaryngol Head Neck Surg. 94 (1986): 143–156

Stammberger H: Functional Endoscopic Sinus Surgery,
Mosby, St. Louis, 1991

TERMINOLOGIA ANATOMICA: International Anatomical Terminology.
Federative Committee on Anatomical Terminology (FCAT).
Thieme Stuttgart. New York 1998

Wigand ME, Steiner W, Jaumann MP:
Endonasal surgery with endoscopical control: from radical operation to rehabilitation of
the mucosa. Endoscopy 10 (1978): 255–260

Wigand ME: Endoscopic Surgery of the Paranasal Sinuses and Anterior Skull Base,
Thieme, Stuttgart, New York,1990

Wigand ME: Transnasal ethmoidectomy under endoscopical control.
Rhinology 19 (1981): 7–15

Wurster CF, Smith DE: The endoscopic approach to the pituitary gland.
Arch Otolaryngol Head Neck Surg 120 (1994): 674

Index

A

abducent nerve (VI)	27, 45, 46
anterior cerebral artery	37, 41, 42, 123, 128
anterior communicating artery (ACoA)	42, 123, 128
arachnoid cyst	X, 121–124

B

basilar artery	20, 21, 51, 52

C

carotid prominence	8, 30, 31, 45, 91–95, 119
cavernous sinus	12, 13, 27, 44, 45, 64, 65
choana	11, 16, 26, 80
clivus	6, 7, 8, 13, 26, 30, 31, 81, 92, 95, 96, 97, 110, 118, 119
craniopharyngeoma	113, 114, 115, 116, 117, 118, 119, 120
crista galli	6, 7, 10
cistern, suprasellar	27, 33, 38, 99, 106, 107, 116, 122, 123, 124

D

diaphragma sellae	27, 36, 38, 39, 40, 107, 108, 120
dorsum sellae	7, 37, 47, 51, 116, 122

E

equipment	VII, XI, XII, XIII, 73, 74, 75, 76, 77
ethmoidal cell	14, 15, 19, 54, 56, 57
EUSTACHIAN tube	13, 17

G

GADOLINIUM	62, 63, 64, 65, 104, 109, 113, 117, 121, 125

H

habenula	49, 50

I

inferior nasal concha	4, 5, 6, 10, 14, 15, 16, 23, 26, 56, 57, 59
instruments	55, 69, 73, 74, 80, 81, 82, 94, 105, 129
internal carotid artery (ICA)	17, 20, 21, 37, 41, 55, 63, 64
interpeduncular cistern	47, 64

M

macroadenoma	X, 64, 65, 104, 105, 106, 107, 1108, 109, 110, 111,

Index

112

maxillary prominence	8, 45
maxillary sinus	10, 14, 15, 56, 57
microadenoma	62
middle nasal concha	4, 6, 10, 14, 15, 26, 29, 54, 56, 57, 59, 80, 84, 86, 87

N

nasal septum	4, 10, 14, 15, 22, 23, 29, 30, 54, 56, 57, 61, 80, 84, 86, 87
nasal spine	4

O

oculomotor nerve (III)	45, 46, 51, 52
ONODI cell	54, 59, 60
ophthalmic artery	27, 44
ophthalmic nerve (VI)	45, 46
optic canal	44, 54
optic chiasm	23, 27, 34–37, 40–43, 55, 128
optic nerve (II)	18, 27, 37, 42, 43, 45, 55, 120, 128
opto-carotid recess	44–46

P

pineal recess	49, 50
pituitary gland	8, 12, 17, 18, 22, 23, 27, 33–39, 41, 55, 62, 81, 82, 99,
	108, 112, 114, 116, 120, 126, 130, 131
pituitary stalk	27, 34, 36, 37, 39, 40, 47–49, 51, 55, 107, 108
sphenoidal plane	30, 31, 38, 44, 90–92, 95, 100, 106, 119, 126, 127
position of the patient	69, 70
posterior cerebral artery	23, 52
posterior communicating artery (PCoA)	51, 52
pterygoidens, canalis	8

R

RATHKE's cleft cyst	X, 125, 126–128

S

sella turcica	IX, 2, 6, 7, 11–13, 16, 17, 26, 27, 30–32, 43–46, 53–54, 55, 58, 59, 61, 63, 69, 80–82, 90–97, 100–102, 108, 110, 112, 116, 119, 120–124, 129, 130

Index

sella turcica, floor of	44, 59, 91–95, 97
sphenoidal sinus ostium	26, 29, 30, 69, 73, 80, 83, 84–90
sphenoidal plane	30, 31, 38, 44, 90–92, 95, 100, 106, 119, 126, 127
sphenoidal rostrum	11, 16, 26, 80, 89
septum of sphenoidal sinus	58, 59, 61, 90, 95, 96
sphenoidal sinus	6–12, 14–17, 22, 23, 26, 27, 29, 30, 54–61, 63–65, 69, 73, 80, 81, 83–92, 95–102, 107, 110, 115, 118, 119, 126, 127, 129
sphenopalatine artery	9
stria medullaris of thalamus	49, 50
superior cerebellar artery (SCA)	20, 23, 51, 52
superior hypophyseal artery	35
superior orbital fissure	27, 45, 46

T

trochlear nerve (IV)	45, 46
third ventricle	28, 47–50, 55, 63

SpringerMedicine

Manfred Tschabitscher, Clemens Klug

Endoscopic Anatomy of the Middle Ear

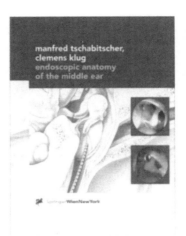

2000. XIII, 127 pages. 246 coloured figures.
Hardcover DM 298,–, öS 2086,–
(recommended retail price)
ISBN 3-211-82973-3

While microsurgical techniques have already become common knowledge in ENT surgery, a new era has begun with the development of endoscopic surgery. Thereby, however, structures on the opposite side are seen best, not those which have already been examined by the endoscope. One solution to this problem is a biportal approach to the middle ear. Filling an essential gap in anatomical and ENT literature, this atlas shows the various approaches to the middle ear, e.g., through the tympanic membrane or through the Eustachian tube, which allow safe surgical manipulations.

Contents
Transmeatal approach
Transtympanic approach
- Anterior inferior transtympanic approach
- Posterior superior transtympanic approach
- Posterior inferior transtympanic approach

Transmastoid approach
- Through antrum
- Through facial recess

Transtubal approach
Two-port approach

Please visit our new website: **www.springer.at**

A-1201 Wien, Sachsenplatz 4–6, P.O. Box 89, Fax +43.1.330 24 26, e-mail: books@springer.at, Internet: **www.springer.at**
D-69126 Heidelberg, Haberstraße 7, Fax +49.6221.345-229, e-mail: orders@springer.de
USA, Secaucus, NJ 07096-2485, P.O. Box 2485, Fax +1.201.348-4505, e-mail: orders@springer-ny.com
Eastern Book Service, Japan, Tokyo 113, 3–13, Hongo 3-chome, Bunkyo-ku, Fax +81.3.38 18 08 64, e-mail: orders@svt-ebs.co.jp

SpringerMedicine

Peter S. Hechl,
Reuben C. Setliff III,
Manfred Tschabitscher

Endoscopic Anatomy of the Paranasal Sinuses

1997. XV, 130 pages. 194 partly coloured figures.
Hardcover DM 298,–, öS 2086,–
(recommended retail price)
ISBN 3-211-82922-9

For the beginner or for the accomplished sinus surgeon, mastering the anatomy of the lateral nasal wall is an ongoing challenge. Even though there are excellent standard anatomical references and equally outstanding sinus courses with cadaver dissection, a reference depicting the surgical anatomy is needed. A step-by-step surgical approach on the anterior nasal spine to the anterior wall of the sphenoid is presented. The sinus surgeon is confronted with a wide range of different spaces created by the ethmoid bone. No other bone in the human body has so many anatomical variations. Four critical anatomical structures are emphasized as the foundation for a precise approach to surgery of the maxillary, anterior ethmoid, frontal, and posterior ethmoid sinuses. The goal of this book is to meet the tremendous challenge of offering an anatomical approach which will serve the sinus surgeon of every level of experience and expertise.

"... an atlas of surgical endonasal anatomy comprising exhaustive illustrations of very high quality. The legends are clearly appended ... The colors and contrasts which are important aids in endoscopy are perfectly presented ... this atlas appears very useful for understanding the anatomy of the ethmoidal labyrinth. It will familiarise the surgeon's eye with the fundamental anatomic landmarks in normal and pathologic conditions ..."

Surgical and Radiologic Anatomy

"... an important aid to the acquisition of endoscopic knowledge and ability ..."

Australian Journal of Otolaryngology

"... a leading anatomical reference book for planning of endoscopic procedures involving the paranasal sinuses and therefore this book should prove useful to ENT-surgeons as well as to neurosurgeons ..."

Minimally Invasive Neurosurgery

SpringerWienNewYork

A-1201 Wien, Sachsenplatz 4–6, P.O. Box 89, Fax +43.1.330 24 26, e-mail: books@springer.at, Internet: www.springer.at
D-69126 Heidelberg, Haberstraße 7, Fax +49.6221.345-229, e-mail: orders@springer.de
USA, Secaucus, NJ 07096-2485, P.O. Box 2485, Fax +1.201.348-4505, e-mail: orders@springer-ny.com
Eastern Book Service, Japan, Tokyo 113, 3–13, Hongo 3-chome, Bunkyo-ku, Fax +81.3.38 18 08 64, e-mail: orders@svt-ebs.co.jp

SpringerMedicine

Henri M. Duvernoy

The Human Brain

Surface, Three-Dimensional Sectional Anatomy with MRI, and Blood Supply

In collaboration with P. Bourgouin, E. A. Cabanis, F. Cattin, J. Guyot, M. T. Iba-Zizen, P. Maeder, B. Parratte, L. Tatu, and F. Vuillier. With drawings by J. L. Vannson
Second, completely revised and enlarged edition
1999. VII, 491 pages. 272 partly coloured figures.
Hardcover DM 398,–, öS 2786,–
(recommended retail price). ISBN 3-211-83158-4

The recent progress of medical imaging due to CT, MRI, and the three-dimensional reconstruction of cerebral structures calls for a better understanding of the anatomy of the brain. Therefore, this book comprises serial sections – 2 mm thick – of the cerebral hemispheres and diencephalon in the coronal, sagittal, and axial planes. So as to point out the level of the sections more accurately, each section is shown from different angles emphasizing the surrounding hemisphere surfaces. This three-dimensional approach has proven to be extremely useful to apprehend the difficult anatomy of the gyri and sulci of the brain.

"This wonderful atlas by Duvernoy and colleagues is a must in every institute dealing with neurological patients and neuroimaging. The atlas is very meticulously made. A correlation is given between MRI and morphology ... The atlas is very good in giving the precise localisation ..."

Clinical Neurology and Neurosurgery

Henri M. Duvernoy
The Human Brain Stem and Cerebellum
Surface, Structure, Vascularization, and Three-Dimensional Sectional Anatomy with MRI

In collaboration with J. F. Bonneville, E. A. Cabanis, F. Cattin, J. Guyot, and M. T. Iba-Zizen.
With drawings by J. L. Vannson. 1995. VII, 430 pages. 168 figures.
Hardcover DM 395,–, öS 2765,– (recommended retail price)
ISBN 3-211-82503-7

"... a thorough and excellent account of the anatomy of the brain stem and cerebellum ... an excellent reference and would be of great value to both trainees and senior surgeons".

British Journal of Neurosurgery

SpringerWienNewYork

A-1201 Wien, Sachsenplatz 4–6, P.O. Box 89, Fax +43.1.330 24 26, e-mail: books@springer.at, Internet: www.springer.at
D-69126 Heidelberg, Haberstraße 7, Fax +49.6221.345-229, e-mail: orders@springer.de
USA, Secaucus, NJ 07096-2485, P.O. Box 2485, Fax +1.201.348-4505, e-mail: orders@springer-ny.com
Eastern Book Service, Japan, Tokyo 113, 3–13, Hongo 3-chome, Bunkyo-ku, Fax +81.3.38 18 08 64, e-mail: orders@svt-ebs.co.jp

Springer-Verlag and the Environment

WE AT SPRINGER-VERLAG FIRMLY BELIEVE THAT AN international science publisher has a special obligation to the environment, and our corporate policies consistently reflect this conviction.

WE ALSO EXPECT OUR BUSINESS PARTNERS – PRINTERS, paper mills, packaging manufacturers, etc. – to commit themselves to using environmentally friendly materials and production processes.

THE PAPER IN THIS BOOK IS MADE FROM NO-CHLORINE pulp and is acid free, in conformance with international standards for paper permanency.

Printed by Publishers' Graphics LLC
DBT130513.15.16.137